STATE OF CALIFORNIA
DEPARTMENT OF NATURAL RESOURCES
GEORGE D. NORDENHOLT, Director

DIVISION OF MINES

WALTER W. BRADLEY
State Mineralogist

GEOLOGIC BRANCH
Ferry Building, San Francisco

OLAF P. JENKINS
Chief Geologist

San Francisco   Bulletin No. 115   October, 1937

BIBLIOGRAPHY

OF THE

GEOLOGY AND MINERAL RESOURCES

OF

CALIFORNIA

For the years 1931 to 1936, inclusive
(Supplementing the Master Bibliography, Bulletin 104)

By

SOLON SHEDD
Stanford University, California

CALIFORNIA STATE PRINTING OFFICE
GEORGE H. MOORE, STATE PRINTER
SACRAMENTO, 1938

51441

LETTER OF TRANSMITTAL

To His Excellency,
THE HONORABLE FRANK F. MERRIAM,
Governor of the State of California.

SIR:

I have the honor to transmit herewith Bulletin No. 115 of the Division of Mines, being the "Bibliography of the Geology and Mineral Resources of California, 1931-1936, Inclusive." Like this division's previous Bulletin No. 104 covering the bibliography to the end of 1930, and which it supplements, this present work has been prepared by Dr. Solon Shedd, librarian of the Branner Memorial (Geological) Library at Stanford University, in cooperation with the Geologic Branch of the Division of Mines, with clerical assistance furnished by the federal Works Progress Administration. As evidenced by the number of entries herein, the past six years have been particularly fruitful in contributions to the literature on the geology and mineral resources of California.

Respectfully submitted.

GEORGE D. NORDENHOLT, Director,
Department of Natural Resources.

November 1, 1937.

TABLE OF CONTENTS

AUTHOR'S PREFACE

PART I (BIBLIOGRAPHY)

The bibliography forming Part I of this publication is intended to include the publications dealing with geology, paleontology, mineralogy, petrology, economic mineral resources and related subjects of California.

An attempt has been made to list all the published material dealing with these subjects that have to do with California in any way from January 1, 1931, to December 31, 1936. This bibliography is also intended to include references to all publications that were omitted from Bulletin 104. In some cases abstracts were published after the articles were published and references made to them in Bulletin 104. Those abstracts have been included in the present bibliography. The plan followed in this bibliography is similar to that used in Bulletin 104 of the Division of Mines.

The entries in this bibliography are made in a definite way. First comes the name of the author, then the title of the article followed by the citation. In most cases the citation is greatly abbreviated and given in a fixed order as follows: The publication, then in the case of serials, the number of the volume followed by a colon and the inclusive pages, after which comes, in parentheses, the year of publication. In case the volume is in a numbered series this number, unless it is the first series, appears in parentheses just before the volume number. When issues are paged separately the serial number is given immediately following the volume number.

The papers of each author are listed in chronologic order by years but no definite order of arrangement is attempted for the papers in any particular year. When papers form a part of a volume that may continue for more than one year, the date on the title page of the volume has been used as the date of issue. Each reference bears an index number, consisting of the last two figures of the year of publication. This is followed by a small letter, should there be more than one publication listed by the same author in any year. These index symbols are made use of in Part II (Index) for the purpose of finding references cited in Part I (Bibliography).

PART II (INDEX)

The Index to the Bibliography, Part II, is arranged alphabetically with cross references according to subject as shown in the title of the article. In the Index, the titles have been very much shortened. The intention has been to give as much information in the Index as possible and make the references easy to find.

In the compilation of this bibliography free use has been made of various bibliographies dealing with geological literature and related subjects, but by far the most helpful have been those prepared by

(vii)

Mr. John M. Nickles. These include those published as bulletins of the U. S. Geological Survey and Annotated Bibliography of Economic Geology of which he has been Editor.

Acknowledgments.

To the Division of Mines, Department of Natural Resources of the State of California, Mr. Walter W. Bradley, State Mineralogist, and Mr. Olaf P. Jenkins, Chief Geologist, I am under obligations for helpful suggestions. To Miss Corinne Kibler, who has acted as assistant in the preparation of this bulletin, I am under obligation for her very careful work at all times. She has been very helpful, not only in copying and checking references, but also in looking through periodicals to find articles dealing with California. To all those persons who were so kind as to take time to check their list of publications, I hereby express my sincere thanks. While it is not possible to mention each one separately from whom help has been received, I wish to say that I appreciate fully any help and suggestions that I have received from any source and hereby acknowledge my gratitude. Clerical assistance was provided by a Federal Works Progress Administration project.

While it is too much to expect a work of this kind to be absolutely perfect the author hopes the number of omissions and mistakes will not be large and that this publication may be of value to persons interested in the geology of California.

Stanford University, California,
June 10, 1937.

SOLON SHEDD.

LIST OF ABBREVIATIONS

abst	abstract
Absts	Abstracts
abt	Abteilung
Ac	academy
Am	American, etc.
An	annals, annual; Annotated
ann	annotation
App	Appendix
Ariz	Arizona
As	association
B	bulletin
Beitr	Beiträge
Bib	bibliography
Biol	biological
Bur	bureau
C R	compte rendu
Cal	California
Chem	chemical, etc.
Circ	circular
Col	collection
Coll	college
Cong	congress
conn	connexes
Contr	contributions
D C	District of Columbia
Dev	development
Div	division
Dom	Dominion
Dp	department
Ec	economic
Ed	edition, editor, etc.
Eng	engineering, engineer
figs.	figures
G	geology, geologist, etc.
G S	geological survey
Geog	geography, etc.
Geol	geological, etc.
Geop	geophysical
Govt	government
Hist	history
I	institute
illus	illustration
incl	including
Inf	information
Inst	institute
Int	international
Invest	investigation
J	journal, etc.
Jb	Jahrbuch
Jr	junior
M	mines, mining
Mag	magazine
Mem	memoirs
Metal	metallurgy
Min	mineral
Miner	mineralogy, etc.
Misc	miscellaneous
Mus	museum
N	new; neues
N H	natural history
n s	new series
Nat	national; natural, etc.
no	number
Obs	observatory
Oc P	occasional papers
Off	office
P	papers
Pacif	Pacific
Pet, Petr	petroleum
pp	pages
Pr	proceedings
Print	printing
Pub	publications
Q	quarterly
R	royal
Ref	referate
Res	research
Rp	report
Rv	review
S	southern, etc.; survey
Sc	science, scientific, etc.
Sci	sciences
Sec	section
Ser	series, serial, etc.
Soc	society
Sp	special
Sta	station
Stat	statistical
Suppl	supplement, etc.
Tech	technology, etc.
Tr	transactions
U S	United States
Univ	university
Wash	Washington
Yb	yearbook
Zool	zoology
Zs	Zeitschrift

SERIALS

Ac N Sc, Phila, Pr. Academy of Natural Sciences, Philadelphia, Proceedings.

Ac Sc, C R. Academie des sciences, Paris, Comptes rendus.

Agr Exp Sta Berkeley Cal, B. Agricultural Experiment Station, Berkeley, California, Bulletin.

Am Ac Arts, Pr. American Academy of Arts and Sciences, Proceedings. Boston.

Am As Petroleum G, B. American Association of Petroleum Geologists, Bulletin.

Am Chem Soc J. American Chemical Society, Journal. New York.

Am G. American Geologist. Minneapolis.

Am Gas J. American Gas Light, Journal. N. Y.

Am Geop Union, Tr. American Geophysical Union, Transactions.

Am I M Eng, Tech Pub, Petr Div, Tr. American Institute of Mining and Metallurgical Engineers, Petroleum Development and Technology, Petroleum Division, Transactions.

Am J Sc. American Journal of Science. New Haven, Conn.

Am J Surgery. American Journal of Surgery.

American Midland Naturalist.

Am Miner J. American Mineralogical Journal. New York.

Am Mineralogist. American Mineralogist.

Am Petroleum Inst, Pr, Sec, Production B. American Petroleum Institute, Proceedings, Section, Development and Production Engineering, Bulletin.

Am Soc Civil Eng, Pr. American Society of Civil Engineers, Proceedings. N. Y.

Am Soc Mechanical Eng, Tr. American Society of Mechanical Engineers, Transactions.

An Bib Ec G. Annotated Bibliography of Economic Geology.

Anal Soc Cientif Argent. Sociedad Cientifica Argentina, Anales.

Aquarium.

Beitr Geoph. Beitrage zur Geophysik. Leipzig.

Biogeographical Soc Japan, B. Biogeographical Society, Japan, Bulletin.

Biol Absts. Biological Abstracts.

Cal Dp Nat Res, Div Mines; M. G, J, St Mineralogist's Rp; Div Mines and Mining, Mining in California. State of California, Department of Natural Resources, Division of Mines, California Journal of Mines and Geology.

Cal Dp Nat Res, Div Oil and Gas, California Oil Fields. State of California, Department of Natural Resources, Division of Oil and Gas.

Cal Dp Pub Works, Cal Highways. State of California, Department of Public Works, California Highways.

Cal Dp Pub Works, Div Water Resources, B. State of California, Department of Public Works, Division of Water Resources, Bulletin.

Cal Oil World. California Oil World. Los Angeles, California.

Cal Univ, Coll Agri, B. California, University of, College of Agriculture, Bulletin.

Cal Univ, Dp G, B. California, University of, Department of Geology, Bulletin.

Cal Univ, Los Angeles, Pub Math & Phys Sc. University of Southern California, Los Angeles, Publication Mathematics and Physical Sciences.

Cal Univ, Pub Zool. California, University of, Publications in Zoology.

Cal Univ, Seism Sta, B. California, University of, Publications Seismographic Stations, Bulletin.

Canada, Dom Obs, Pub, Bib Seism. Publications of the Dominion Observatory, Ottawa, Canada, Bibliography of Seismology.

Carnegie Inst, Wash, News Service B, School Ed. Carnegie Institution of Washington (D. C.) News Service Bulletin, School Edition.

Chem Absts. Chemical Abstracts.

Chem Met Eng. Chemical and Metallurgical Engineering.

Chemicals.

Chimie et Industrie.

Civil Eng. Civil Engineering.

Compressed Air.

Condor.

Congrés intern Mines Met et Geol appl. Congrés international des mines, de la metallurgie, de la mechanique et de la géologie appliquées.

Cushman Lab Foram Res, Contr. Contributions from the Cushman Laboratory for Foraminiferal Research.

Deutsche Rundschau. Berlin.

SERIALS—Continued

(The abbreviation used in the citation is printed in black-faced type.)

Earthquake Res Inst, B. Earthquake Research Institution, Bulletin. Tokyo Imperial University.

Ec G. Economic Geology. Lancaster, Pa.

Eng Index. Engineering Index.

Eng M J. Engineering and Mining Journal.

Eng News-Rec. Engineering News-Record.

Entomological News.

Ergebnisse der Kosmischen Physik.

Explosives Eng. Explosives Engineer.

Far-East Geological and Prospecting Trust of U.S.S.R.

G Rundschau. Geologische Rundschau. Leipzig.

G Soc Am, B, Pr. Sp P. Geological Society of America, Bulletin, Proceedings, Special Paper.

G Zentralbl. Geologisches Zentralblatt. Anzeiger fur Geologie, Petrographie, Palaeontologie, und verwandte Wissenschaften, Leipzig.

Gas Age-Rec. Gas Age-Record.

Geog J. The Geographical Journal.

Geog Rv. The Geographical Review.

Geog Soc Philadelphia, B. Geographical Society of Philadelphia, Bulletin.

Inst Pet Tech, J. Institute of Petroleum Technology, Journal.

Int G Cong, Guide Book. International Geological Congress, Guide Book.

J Botany. Journal of Botany.

J G. Journal of Geology. Chicago, Ill.

J Mammalogy. Journal of Mammalogy.

J Paleontology. Journal of Paleontology.

J Sed Petrology. Journal of Sedimentary Petrology.

Kyoto Imperial Univ. Kyoto Imperial University.

Los Angeles Jr Coll, Pub. Los Angeles Junior College, Publications.

Los Angeles Museum, Pub. Los Angeles Museum of History, Science, and Art, Publications.

M J Ariz. Mining Journal, Arizona.

M Mag. Mining Magazine; later Mining and Statistic Magazine; Mining Magazine and Journal of Geology, Mineralogy, Metallurgy, Chemistry and the Arts. N. Y.

M Metal. Mining and Metallurgy.

M Sc Press. Mining and Scientific Press.

Madroño, Cal Botanical Society. Madroño, California Botanical Society journal.

Military Eng. Military Engineer.

Min Absts. Mineral Abstracts.

Min Ind. Mineral Industry.

Min. Res. Mineral Resources.

Miner Mag. Mineralogical Magazine and Journal of the Mineralogical Society. London.

Miner Soc S Calif, B. Mineralogical Society of Southern California, Bulletin.

Mineralogist.

Mining Rv. Mining Review.

Musee Royal d'Histoire Naturelle de Belgique.

N Jb. Neues Jahrbuch fur Mineralogie, Geologie, und Paläeontologie.

Nat Ac Sc, Pr. National Academy of Sciences, Proceedings.

Nat Geog Soc, Mag. National Geographic Society Magazine.

Nat Hist. Natural History.

Nat Mus Canada, B. National Museum of Canada, Bulletin.

Nat Pet News. National Petroleum News. Cleveland, Ohio.

Nat Res Council, B, Rp Committee Sedimentation, Circ Ser. National Research Council, Bulletin, Report Committee Sedimentation, Circular Series.

Nature. London.

Nautilus.

Oil B. Oil Bulletin.

Oil Gas J. Oil and Gas Journal.

Pacif Hist Rv. Pacific Historical Review.

Pacific Dental Gazette.

Pacific M News. Pacific Mining News.

Pacific Mineralogist.

Pale Zentralbl. Palaeontologisches Zentralblatt.

Pan Am G. Pan American Geologist. Des Moines, Iowa.

Peabody Mus N H. Peabody Museum of Natural History.

Petermann's Mitt. Petermann's Mitteilungen aus Justus Perthes Geographischer. Anstalt, Gotha.

Pet Tech, J. Petroleum Technology Journal.

Petroleum Times.

Petroleum World, An Rv. Petroleum World, Annual Review, Los Angeles, California.

Petroleum World, London.

Physics.

Physikalische Zs. Physikalische Zeitschrift.

Pit and Quarry.

Pop Sc Mo. Popular Science Monthly.

Q J G Soc London. Quarterly Journal of the Geological Society of London.

R Soc London, Pr. Royal Society of London, Proceedings.

Revue des Produits Chimiques.

Rocks & Min. Rocks and Minerals.

Rv Geol et Sci conn. Revue de Geologie et des Sciences connexes.

Rv Univ Mines. Revue universelle des mines. Liége and Paris.

S Cal Ac Sc, B. Southern California Academy of Sciences, Bulletin.

SERIALS—Continued

(The abbreviation used in the citation is printed in black-faced type.)

San Diego Soc N H, Tr. San Diego Society of Natural History, Transactions.

San Diego State Coll, Delvers Quarterly. San Diego State College, Delvers Quarterly.

Santa Barbara Mus N H, Oc P. Santa Barbara Museum of Natural History, Occasional Papers.

Sc Mo. Scientific Monthly. N. Y.

Sc Progress. Science Progress.

Science ns. Science, new series. N. Y.

Scripps Inst Oceanography, B. Scripps Institute of Oceanography, Bulletin.

Seism Soc Am, B, Eastern Sec, Earthquake Notes. Seismological Society of America, Bulletin, Eastern Section, Earthquake Notes.

Smiths Misc Col, An Rp. Smithsonian Miscellaneous Collections, Annual Report.

Soc G Belgique, An. Société géologique de Belgique, Annales. Liége.

Standard Oil B. Standard Oil Bulletin.

Stanford Univ Micropaleontology B. Stanford University Micropaleontology Bulletin.

Stanford Univ Press. Stanford University Press.

Terre Vie. Terre et la Vie.

Terrestrial Magnetism and Atmospheric Electricity.

Torrey Botanical Club, B. Torrey Botanical Club, Bulletin.

U S B M, Min Yb, Min Res, Calendar Year, Inf Circ, List Pub, Stat. App, Rp Invest, Ec P, Geophysical Absts, Tech P, B. United States Bureau of Mines, Minerals Yearbook, Mineral Resources, Information Circular, List of Publications, Statistical Appendix, Report of Investigation, Economic Paper, Geophysical Abstracts, Technical Paper, Bulletin.

U S Bur Rec. United States Bureau of Reclamation.

U S Coast S, Sp Pub, Serial. United States Coast and Geodetic Survey, Special Publication.

U S Dp Interior. United States Department of the Interior.

U S G S, Sp Pub, B, P.P, W-S P, Type-Script Rp, Circ, Monographs, An Rp. United States Geological Survey, Special Publication, Bulletin, Professional Paper, Water-Supply Paper, Typescript Report, Circular, Annual Report.

U S Nat Mus. Pr. United States National Museum, Proceedings.

Union Oil Company, B. Union Oil Company, Bulletin.

Univ Mexico, B. University of Mexico, Bulletin.

Volcano Letter.

Wash Ac Sc, J. Washington (D.C.) Academy of Sciences, Journal.

Western City.

Western Miner & Prospector.

Westways.

Yosemite Nature Notes.

Zoolbot Gesell. Zoologisch-botanische gesellschaft.

Zs Geom., Zeitschrift für Geomorphologie.

Zs Gletscherk. Zeitschrift für Gletscherkunde, Berlin.

Zs Prak G. Zeitschrift für praktische Geologie. Berlin.

Zs Vulkan. Zeitschrift für Vulkanologie.

ERRORS IN BULLETIN No. 104

Page   6—Lieutenant R. S. Williamson. In the first line "set" should be "sent".

Page   45—Bealey, A.—52
        Chem Soc Q.J. 4:180 (1852) should read Chem Soc Q.J. 4:180-186 (1852)

Page   47—Berkey, Charles P.—29a
        Berry, Edward W. is the correct author for this article, not Berkey.

Page   47—Bertrand, Emile
        Reference 81 should be 78.
        Spelling error—Kyrst should be Kryst.

Page   49—Blake, William Phipps
        references 55a and 55g are the same, with the correct title under 55g, and
        the correct references under 55a. Reference 64 should refer to vol. 9,
        not vol. 10.

Page   88—Dolbear, Samuel H.—15a
        (1950) should be (1915)

Page 113—Guild, F. N.
        Reference 11 is merely a part of 11a.
        The inclusive paging is 321-331.

Page 128—Hofmann, Dr.—49
        Should be Hofman, Dr.

Page 134—Irelan, William, Jr.—93
        The last two lines should read Wm. G. Hodson, R. L. Dunn and Harold W.
        Fairbanks.

Page 135—Jannetaz, Edouard—86
        This reference should not appear in Bulletin 104 because it deals with
        material from Boleo, Lower California. There is no mention of anything
        from the State of California.

Page 137—Johnson, Harry Roland—27
        Should read Division of Water Rights, not Div Eng and Irrigation.

Page 145—Koechlin, R.—07
        This reference should not be in No. 104 as it is merely a list similar to
        that of L. J. Spencer.

Page 148—Lahmann, W. M.
        Should be Lehmann, W. M.

Page 159—Louderback, George Davis—08a
        Title is incorrect. It should be "Benitoite, its paragenesis and mode of
        occurrence".
        Reference 08b should be to vol. 18 of G Soc Am, B, not to 19.

Page 196—Papish, Jacob—29
        This paper is by Papish and Holt only. Reference should also be made
        to the original of this article which appeared in the Journal of Physical
        Chemistry, vol. 32, no 1:142-147, 2 tables (1928).
        Brewer's name should also be eliminated from page 130—Holt, Donald A.
        —29, and his name and the reference also eliminated from page 56.

Page 202—Price, T.—83
        An Rp 3:311-312 (1883) should be An Rp 3:86-87 (1883)

Page 206—Reed, R. D.—28a
        The reference should read as follows: The occurrence of feldspar in
        California sandstone. Discussion. Am As Petroleum G, B 12:1023-1024
        (1928)

Page 329—At top of page.
        Mono—Cont. should read Mono County—Cont.

Page 359—At top of page, right hand column.
        Should read Sierra Nevada, southern—Cont.

(xiv)

BIBLIOGRAPHY OF THE GEOLOGY AND MINERAL RESOURCES OF CALIFORNIA

1931 TO 1936 (INCLUSIVE)

By SOLON SHEDD

PART I

BIBLIOGRAPHY

Anonymous

31 New surveys made by magneto-meter. (Editorial note) Cal Oil World 23:15 (1931) (*abst*) Rv Geol et Sci conn 12:153-154 (1932)

32 Three dams on San Andreas fault have resisted earthquakes. Eng News-Rec 109:218-319 (1932) (*abst*) Canada, Dom Obs, Pub, Bib Seism 10:269 (1933)

33 Borates (second edition 1920-1932): the mineral industry of the British Empire and foreign countries. (Great Britain), Imperial Institute, 44 pp London (1933)

33a Hazardous tunneling at Hetch Hetchy. Eng News-Rec 110, no 22:701-704 (1933) (*ann*) An Bib Ec G 6:111 (1934)

34 Debris flow from canyons in Los Angeles County flood. Eng News-Rec 111:439-440 (1934) (*abst*) Rv Geol et Sci conn 14:457 (1934)

Abozeid, Mahmood

31 Microscopic study of California oil field emulsions. Am I M Eng, Petr Div, Tr 1931:340-358 (1931) (*ann*) An Bib Ec G 5:133 (1932) (*abst*) M Metal 11:28 (1930)

Adams, Bradford C.

32 An ecologic analysis of a Pliocene faunule from southern California. Stanford Univ Micropaleontology B 3:122-127 (1932)

35 (and **Cushman,** J. A.) Two late Tertiary *Bolivinas* from California. Cushman Lab Foram Res, Contr 11:16-20, figs (1935)

Adams, Oscar S.

30 The Bowie method of triangulation adjustment. U S Coast S, Sp Pub 159:32 pp (1930)

Adams, W. W.

32 (and **Galiher,** C.) Fuller's earth. U S B M, Min Yb 1932-1933:709-714, California 709, 711, 712, 713 (1933)

Adams, W. W.—Cont.

34 (and **Metcalf,** R. W.) Fuller's earth. U S B M, Min Yb 1934:969-974, California 971, 974 (1934)... 1935:1063-1068 (1935)... 1936:947-952, California 950, 951 (1936)

Aitkens, Irene

31 Feldspar gems (Amazon stone, Moonstone, Sunstone). U S B M, Inf Circ 6533:1-11, California 6, 8-9 (1931) (*abst*) U S B M, List Pub 1910-1932:152 (1933) (*abst*) Eng Index 1931:1268 (1931)

31a Garnets (gem stones). U S B M, Inf Circ 6518:1-12, California 4, 7 (1931) (*abst*) U S B M, List Pub 1910-1932:151 (1933)

31b Opals. U S B M, Inf Circ 6493:9 pp (1931) (*abst*) U S B M, List Pub 1910-1932:150 (1933)

31c Turquoise. U S B M, Inf Circ 6491:1-18, California 10 (1931) (*abst*) U S B M, List Pub 1910-1932:150 (1933)

31d Topaz. U S B M, Inf Circ 6502:11 pp (1931) (*abst*) U S B M, List Pub 1910-1932:150 (1933)

31e Tourmaline. U S B M, Inf Circ 6359:1-9, California 5-6 (1931) (*abst*) Eng Index 1931:1268 (1931)

32 Quartz gem stones. U S B M, Inf Circ 6561:15 pp (1932) (*abst*) U S B M, List Pub 1910-1932:154 (1932)

Allan, M.

32 (and **Hughes,** H. H.) Sand and gravel. U S B M, Stat App, Min Yb 1932-1933:289-297, California 293, 294 (1934)... 1935:939-948, California 942 (1935)

Allen, A. W. (Ed)

30 Vertical range of gold deposits in California. Eng M J 130:210-211 (1930)

31 Geology of the Yosemite Valley. Eng M J 131:204 (1931)

Allen, E. T.

30 (and **Day,** A. L.) The volcanic activity and hot springs of Lassen Peak, California. The Volcano Letter no. 293:

Allen, E. T.—Cont.

1-3 (1930) (*abst*) Zs Vulkan 9:271-278 (1925-1926)

Allen, M. W.

34 (**Wood, H. O.**; and **Heck, N. H.**) Destructive and near-destructive earthquakes in California and western Nevada, 1769-1933. U S G S, Pub 191:24 pp (1934)

Allen, R. C.

34 (and **Campbell, A. B.**) Story of Grass Valley—past and present. M J, Ariz 17, no 18:3-4, 24 (1934) (*abst*) Eng Index 1934:536 (1934)

Allen, Robert E.

30 The mechanics of crude oil curtailment in California. Petroleum World 28, no 11:52 (1930)

31 Control of California oil curtailment. *In* Petroleum Dev and Tech, Am I M Eng, Tr 1931:47-65 (1931)

31a Some problems of oil proration. Petroleum World 28, no 12:37 (1931)

32 How the umpire fixes your potential. Petroleum World 29, no 1:23 (1932)

33 Theory and practice of directed drilling. Am I M Eng, Tr 107:34-41 (1934) Petroleum World 30, no 10:8 (1933) M Metal 14:501-504 (1933)

33a Voluntary oil curtailment in California. Petroleum World An Rv 1933:135 (1933)

34 Some effects of curtailment on the potential and recovery of petroleum in California. M Metal 15:486-489, 4 figs (1934)

35 (and **Pemberton, J. R.**) How much oil can California produce? Petroleum World 32, no 11:94 (1935)

Allen, Victor T.

31 The Ione formation of California. (*abst*) N Jb 1931:225-227 (1931) (*abst*) G Zentralbl 44:60 (1931) (*Review*) J G 39:693-694 (1931) (*abst*) Zs Geom 7:268-269 (1932)

Allison, Ira S.

35 Late Pleistocene topographic correlations on Pacific Coast. (*abst*) Pan Am G 63:310 (1935)

Allyne, A. B.

32 Notes on a distribution soil survey. Gas Age-Rec 70:207-208, 219 (1932)

American Institute of Mining and Metallurgical Engineers, Petroleum Division.

33 Petroleum developments and technology, 1933. Am I M Eng, Tr 103:426 pp (1933)

Anderson, Charles A.

27 Voltaite from Jerome, Arizona. Am Mineralogist 12:287-290 (1927)

30 (and **Knopf, A.**) The Engels copper deposits, California. (*ann*) An Bib Ec G 3:60 (1931) (*abst*) Eng Index 1930:450 (1930) (*abst*) N Jb 1931, Ref 2:439-440 (1931) (*abst*) Zs Prak G 39:30 (1931)

30a Opal stalactites and stalagmites from a lava tube in northern California. (*abst*) N Jb 1931, Ref 1:131 (1931) (*abst*) G Soc Am, B 42:310 (1931) (*abst*) Zs Vulkan 14:92 (1931) (*abst*) G Zentralbl 45:294 (1931)

31 The geology of the Engels and Superior mines, Plumas County, California, with a note on the ore deposits of the Superior mine. Cal Univ, Dp G, B 20:293-330, pls 51-54, 3 figs (1931) (*ann*) An Bib Ec G 4:40 (1931) (*abst*) G Zentralbl, Abt A 48:220-221 (1932)

31a (and **Finch, R. H.**) Quartz-basalt eruptions of Cinder Cone. (*abst*) Zs Vulkan 14:94-95 (1931) G Soc Am, B 41:157 (1930) (*abst*) N Jb 1931, Ref 2:277 (1931) (*abst*) G Zentralbl 45:488-489 (1931)... 46:232-233 (1932)

33 Volcanic history of Glass Mountain, northern California. Am J Sc 26:485-506 (1933) Chem Absts 28:729 (1934)

33a The Tuscan formation of northern California. Cal Univ, Dp G, B 23:215-276, 5 pls, 13 figs (1933) (*Review*) Sc Progress 29:121 (1934) (*abst*) Zs Vulkan 16:281 (1936)

34 Alteration of lavas surrounding hot springs in Lassen Volcanic National Park. Am Mineralogist 20:240-252, 2 tables (1935) (*abst*) Pan Am G 61:373 (1934) (*abst*) G Soc Am, Pr 1934:330-331 (1935) (*abst*) Rv Geol et Sci conn 15:489 (1935) (*abst*) Eng Index 1935:501 (1935) Chem Absts 29:6544 (1935) (*abst*) G Zentralbl, Abt A, Bd 57:36 1936)

35 Volcanic history of Clear Lake area. (*abst*) Pan Am G 64:66-67 (1935) (*abst*) G Soc Am, Pr 1935:343-344 (1936) G Soc Am, B 47:629-664 (1936)

Anderson, Frank Marion

31 Kreyenhagen shales and the Lillis shale. (*abst*) G Soc Am, B 42:302-303 (1931)

31a Upper Cretaceous (Chico) deposits in Siskiyou County, California. Cal Dp Nat Res, Div Mines and Mining, Mining in California, St Mineralogist's Rp 27:11-14, 2 figs (1931)

31b The genus *Fagesia* in the upper Cretaceous of the Pacific Coast. J Paleontology 5:121-126, 1 fig, 3 pls (1931) (*abst*) Pale Zentralbl 1:54 (1932)

31c Age of Horsetown beds of California. (*abst*) G Soc Am, B 42:317-318 (1931)

Anderson, Frank Marion—Cont.

32 Jurassic and Cretaceous divisions in the Knoxville-Shasta succession of California. Cal Dp Nat Res, Div Mines and Mining, Mining in California, St Mineralogist's Rp 28:311-328 (1932)

33 Knoxville-Shasta succession in California. G Soc Am, B 44:1237-1270, 3 pls (1933) Biol Absts 9:2169, entry 19682 (1935) (*abst*) Pale Zentralbl 5:88 (1934) (*abst*) Rv Geol et Sci conn 15:381-382 (1935)

33a Type area of Jurassic Knoxvillian series of California. Pan Am G 60:175-188 (1933)

Anderson, George Harold

31 Geology of the northern half of the White Mountain quadrangle of California-Nevada. (*abst*) G Soc Am, B 42:312-313 (1931)

33 Pseudo-cataclastic texture of replacement origin. (*abst*) Pan Am G 59: 317-318 (1933) (*abst*) G Soc Am, Pr 1933: 316 (1934)

34 (and **Maxson,** J. H.) Physiography of northern Inyo Range (*abst*) Pan Am G 61:314-315 (1934) (*abst*) G Soc Am, Pr 1934:318 (1935)

34a Pseudo-cataclastic texture of replacement origin in igneous rocks. Am Mineralogist 19:185-193 (1934)

35 Granitization and albitization in Inyo Range. (*abst*) Pan Am G 64:66 (1935) (*abst*) G Soc 1935:63 (1936) (*abst*) G Soc Am, Pr 1935:343 (1936)

35a (and **Maxson,** J. H.) Terminology of surface forms of the erosion cycle. J G 43:88-96 (1935)

Anderson, Robert

33 The diatomaceous and fish-bearing Beida stage of Algeria. J G 41:673-698, California 685, 693-695 (1933)

Annis, Wilbert

36 (and **Byerly,** P.) Earthquakes in northern California and the registration of earthquakes at Berkeley, Mount Hamilton, Palo Alto, San Francisco, Ferndale from July 1, 1935 to September 30, 1935. Cal Univ Seism Sta B 5:40-78 (1936)... October 1, 1935 to December 31, 1935, B 5:80-117 (1936)

36a (and **Wilson,** J. T) Earthquakes in northern California and the registration of earthquakes at Berkeley, Mount Hamilton, Palo Alto, San Francisco, Ferndale from April 1, 1935 to June 30, 1935. Cal Univ Seism Sta, B 5:1-38 (1936)

Antonius, Otto

33 Ueber einen Pferdeschadel aus dem Rancho La Brea. Zoolbot Gesell, Band 83, Heft 3-4 pp 39-40 (1933)

Applin, E. R.

33 A microfossiliferous upper Cretaceous section from South Dakota. J Paleontology 7:215-220 (1933)

34 Correlation of California *Turritella andersoni* zone with Gulf Coast Eocene. (*abst*) Pan Am G 62:78 (1934) (*abst*) G Soc Am, Pr 1934:392 (1934)

Arnold, Ralph

27 (and **Loel,** W.) New oil fields of the Los Angeles basin, California. (*abst*) N Jb 1927, Ref 2, Abt B:113 (1927)

31 (and **Kemnitzer,** W. J.) Petroleum in the United States and possessions. 1052 pp, maps, charts, tables (1931)

Ashauer, Hans

36 (**Hollister,** J. S., and **Reed,** R. D.) Sedimentation und faltung im sudlichen Kalifornien. Stille-festschrift 1936:232-233 (1936)

Ashley, James F.

34 (and **Miller,** A. H.) Goose footprints on a Pliocene mud flat. Condor 36:178-179, 1 fig (1934) (*abst*) U S G S, B 869:165 (1933 and 1934)

Atwill, E. R.

31 Truncation of Maricopa sandstone members, Maricopa Flat, Kern County, California. Am As Petroleum G, B 15: 671-688, 5 figs (1931) (*abst*) Rv Geol et Sci conn 12:465 (1932) (*abst*) G Zentralbl 45:307 (1931) (*abst*) Inst Pet Tech, J 17: 423A-424A (1931) (*abst*) Eng Index 1931: 1036 (1931) (*abst*) N Jb 1932, Ref 2:114 (1932)

35 Oligocene Tumey formation of California. Am As Petroleum G, B 19: 1192-1204, figs (1935) (*abst*) Rv Geol et Sci conn 15:383-384 (1935) (*abst*) Eng Index 1935:501 (1935) (*ann*) An Bib Ec G 8:358 (1936)

Averill, Charles Volney

31 Preliminary report on economic geology of the Shasta quadrangle. Cal Dp Nat Res, Div Mines and Mining, Mining in California, St Mineralogist's Rp 27:3-65, illus, map (1931) (*ann*) An Bib Ec G 4:10 (1931) (*abst*) G Zentralbl, Abt A 48:245 (1932) (*abst*) Eng Index 1931:898 (1931)

31a The Mountain Copper Company Ltd. Cyanide treatment of gossan. Cal Dp Nat Res, Div Mines and Mining, Mining in California, St Mineralogist's Rp 27:129-139 (1931)

33 Gold deposits of the Redding and Weaverville quadrangles. Cal Dp Nat Res, Div Mines, M G J, St Mineralogist's Rp 29:2-73, 18 figs, 3 pls (incl geol map) (1933) (*abst*) Eng Index 1934:533 (1934)

Averill, Charles Volney—Cont.

(ann) An Bib Ec G 6:223 (1934) (abst) G Zentralbl, Abt A, Bd 53:330 (1934)

33a The Shasta County copper belt, California, Copper Resources of the World: International Geological Congress, Washington XVI:237-240 (1933)

34 Current mining developments in northern California. Cal Dp Nat Res, Div Mines, M G, J, St Mineralogist's Rp 30:303-309 (1934)

34a Northernmost gold producers led by Mountain Copper. Eng M J 135:516-517 (1934)

35 Mines and mineral resources of Siskiyou County. Cal Dp Nat Res, Div Mines, M G, J, St Mineralogist's Rp 31: 255-338, 13 figs, 1 pl (county map) (1935)

36 (and Erwin, Homer D.) Mineral resources of Lassen County. Cal Dp Nat Res, Div Mines, St Mineralogist's Rp 32:405-444 (1936)

36a Mineral resources of Modoc County. Cal Dp Nat Res, Div Mines, St Mineralogist's Rp 32:445-457 (1936)

Axelrod, D. I.

34 A Pliocene flora from the Eden beds. Am Mus Novitates 729:4 (1934) (abst) N Jb 1935, Ref 3:151 (1935)

Ayers, W. O.

14 Borax. Pop Sci Mo 21:350-361 (1882) (abst) Min Res 1913, Part 2:526 (1914)

Backus, H.

32 (and Hopkins, G. R.) Natural gas. U S B M Min Res, Calendar Year 1929: 319-340, California 319, 320, 322, 324, 325, 326, 327, 328, 330, 331, 332, 334, 336, 337, 339, 340 (1932)... 1930:457-481, California 459, 461, 462, 465, 466, 468, 471, 472, 475, 477, 478, 480, 481 (1932)... 1931:349-372, California 351, 352, 358, 360, 363, 366, 369, 372 (1933)... 1932-1933 Min Yb, Stat App:103-116, California 104, 105, 107, 108, 112, 115 (1934)... 1933:121-135, California 122, 123, 127, 128, 132, 134 (1934)... 1934:31-45, California 32, 33, 36, 39, 42, 44 (1935)

Bacon, Charles S. Jr.

31 Chemistry of the California igneous rocks graphically represented by Niggli-Becke method. (abst) G Soc Am, B 43: 232 (1932) (abst) Pan Am G 55:368 (1931)

33 Geology of Riverside area, California. (abst) Pan Am G 59:313-314 (1933) (abst) G Soc Am, Pr 1933:311-312 (1934)

Bagg, R. M.

08 Miocene foraminifera from the Monterey shale of California, with a few species from the Tejon formation. (abst) N Jb 1908, Band 2:283 (1908)

Bagley, B. W.

32 Cement. U S B M, Min Res, Calendar Year 1929:389-340, California 390, 393, 394, 396, 397, 398, 401, 402, 403, 404, 405, 406, 411 (1933)... 1930:397-432, California 398, 401, 407, 408, 410, 411, 412, 415, 416, 418, 419, 423 (1933)... 1931:523-552, California 526, 527, 529, 530, 531, 534, 535, 536, 537, 538, 542 (1934) Stat App Min Yb 1932-1933:481:514, California 482, 483, 489, 490, 491, 497, 498, 499, 500, 502, 504, 505 (1934)... 1934:65-92, California 67, 68, 70, 71, 72, 74, 76, 78, 79, 81, 84 (1935)... Min Yb 1936:789-807, California 791, 792, 794, 795, 796, 798 (1936)

32a (and Hughes, H. H.) Cement. U S B M, Min Yb 1932-1933: 565-575, California 571 (1933)... 1934:775-797, California 782 (1934)

35 (Hughes, H. H.; and Shuey, E. T.) Cement. U S B M, Min Yb 1935:883-909, California 887, 898, 901, 904 (1935)

Bailey, Thomas L.

31 Eocene age of the Markley formation. (abst) G Soc Am, B 42:304 (1931)

31a The geology of the Potrero Hills and Vacaville region, Solano County, California. Cal Univ, Dp G, B 19:321-333, 2 pls (1931) (abst) N Jb 1933, Ref 3:558-559 (1933)

34 Lower Pleistocene lateral change of fauna near Ventura. G Soc Am, B 46: 489-502, 1 pl, 1 fig (1935) (abst) G Soc Am, Pr 1934:312 (1935) (abst) Pan Am G 61:309 (1934) (abst) Rv Geol et Sci conn 16: 114 (1936)

35 (and Morse, R. R.) Geological observations in the Petaluma district, California. G Soc Am, B 46:1437-1456, 2 figs, pl (1935)

35a Lateral change of fauna in the lower Pleistocene. G Soc Am, B 46:489-502, 1 pl, 1 fig (1935) (abst) Rv Geol et Sci conn 16:114 (1936)

Baily, Joshua L. Jr.

34 The first Pacific conchologist. Nautilus 48:73-76 (1935)

Baker, Frank C.

34 New Lymnaeidae from the United States and California: I. California, Oregon and other western states. Nautilus 48: 17-20 (1934)

36 The freshwater mollusk Helisoma corpulentum and its relatives in Canada. Nat Mus Canada, B 79:1-37 (1936)

Balk, R.

31 (and Cloos, E.) Primary structure of the Sierra Nevada intrusive in a cross section from Yosemite Valley to Mono Lake, California. (abst) N Jb 1931, Ref 2:311 (1931)

Balk, R.—Cont.

36 Structure elements of domes. Am As Petroleum G, B 20:51-67, California 53, 56, 57, 66-67 (1936)

Ball, Sydney H.

33 Diamond deposits of magmatic origin. *In* Ore deposits of the western states. Am I M Eng, Pub 524:526 (1933)

35 Precious and semi-precious stones (gem minerals). U S B M, Min Yb 1935: 1193-1212, California 1194, 1195, 1196 (1935)

36 Gem stones, U S B M, Min Yb 1936: 1051-1056, California 1052 (1936)

Ballard, J. I.

33 Building damage sustained in California earthquake. Eng News-Rec 110: 378-381, 9 figs (1933) (*abst*) Canada, Dom Obs, Pub, Bib Seism 10:305 (1933)

Banks, D. M.

33 (and **Bowles**, O.) Onyx marble and travertine. U S B M, Inf Circ 6751:1-11, California 5 (1933)

36 (and **Bowles**, O.) Lime. U S B M, Inf Circ 6884:37 pp, California 11, 21, 24 (1936)

Barbat, William F.

30 Notes on subsurface methods employed in parts of the San Joaquin Valley. Stanford Univ, Micropaleontology B 2:1-3 (1930)

31 (and **Weymouth**, A. A.) Stratigraphy of *Borophagus littoralis* locality, California. Cal Univ Dp G, B 21, no 3:25-36, 2 figs, 2 pls (1931) (*abst*) G Zentralbl Abt A 48:253 (1932) (*abst*) Pale Zentralbl 2: 266-267 (1933)

32 (and **Cunningham**, G. M.) Age of producing horizon at Kettleman Hills, California. Am As Petroleum G, B 16: 417-421 (1932) (correction)... 16:611-612 (1932) (*ann*) An Bib Ec G 5:145 (1932) (*abst*) Eng Index 1932:952 (1932)

32a (and **Cushman**, J. A.) Notes on some arenaceous foraminifera from the Temblor formation of California. Cushman Lab, Foram Res, Contr 8:29-40 (1932) (*abst*) Rv Geol et Sci conn 13:189 (1933) Pale Zentralbl 2:353-354 (1933)

33 (and **von Estorff**, F.) Lower Miocene foraminifera from the southern San Joaquin Valley, California. J Paleontology 7:164-175, 1 pl, 1 fig (1933) (*abst*) N Jb 1934, Ref 3:422 (1934) Pale Zentralbl 5:271-272 (1934) Biol Absts 9:916, entry 8195 (1935)

34 (and **Galloway**, J.) San Joaquin clay, California. Am As Petroleum G, B 18:476-499 (1934) (*abst*) Inst Pet Tech, J 20:347A-348A (1934) Rv Geol et Sci conn 14:330-331 (1934) (*ann*) An Bib Ec G 7:98 (1935) (*abst*) Eng Index 1934:802 (1934)

Barbat, Wiliam F.—Cont.

34a (and **Johnson**, F. L.) Stratigraphy and foraminifera of the Reef Ridge shale, upper Miocene, California. J Paleontology 8:3-17, 1 pl (1934) (*abst*) Pan Am G 59:239 (1933) (*abst*) N Jb 1934, Ref 3:624 (1934) Biol Absts 9:1396, entry 12523 (1935)

Barber, Ray J.

36 Scott River lifts itself for hydraulic mining. Eng M J 137:171-174 (1936)

Barbour, Percy E.

31 Copper. Min Ind 39:116-212, California 134, 145 (1931)... 40:107-163, California 131 (1932)... 41:107:131, California 109, 122, 131 (1933)... 42:113-148, California 117 (1934)... 43:116-174, California 121 (1935)

Barksdale, Julian D.

32 (and **Cushman**, J. A.) Eocene foraminifera from Martinez, California. (*abst*) Rv Geol et Sci conn 12:632-633 (1932) Nature 128:1080 (1931) (*abst*) Pale Zentralbl 2:291 (1933)

Barnes, R. M.

30 (and **Bowes**, G. H.) Seal Beach oil field. Cal Dp Nat Res, Div Oil and Gas, California Oil Fields 16, no 2:9-31, 9 figs, 8 pls (1930) (*abst*) Rv Geol et Sci conn 13:234 (1933) (*abst*) Eng Index 1932:899 (1932)

32 (and **Bell**, A. H.) Proration on the basis of uniform allowable gas production. *In* Petroleum development and technology. Am I M Eng, Tr 103:142-147 (1932)

Barton, Cecil L.

31 A report on Playa del Rey oil field (Los Angeles County) California. Cal Dp Nat Res, Div Oil and Gas, California Oil Fields 17, no 2:5-15, 5 pls (1931) (*abst*) Rv Geol et Sci conn 13:315 (1933) (*abst*) Eng Index 1933:774 (1933)

Baskerville, Charles

04 (and **Kunz**, G. F.) Kunzite and its unique properties. Am J Sc (4) 18:25-28 (1904)

Baumhauer, H.

99 Ueber sogennante anomale Aetzfiguren an monoklinen Krystallen, insbesondere am Colemanit. Zs Kryst 30:97-117 (1899)

Beatty, M. E.

33 Mountain sheep found in Lyell Glacier. Yosemite Nature Notes 12, no 2:110-112, 1 fig (1933) Biol Absts 8:2503, entry 21473 (1934)

Becker, G. F.

90 Geology of the quicksilver deposits of the Pacific slope. U S G S, Monograph XII:486* pp (1888) (*abst*) Am G 5:178-180 (1890) Am Nat 24:850-851 (1890) Am J Sc (3) 39:68-69 (1890) Eng M J 49:137-138 (1890)

Beebe, Jas. W.

35 History, geology of Dudley Ridge. Cal Oil World 28, no 16:72 (1935)

Beechey, F. W.,

39 The zoology of Captain Beechey's voyage. 180 pp., 50 pls, California 174 (1839)

Behre, C. H. Jr.

33 (and Loughlin, G. F.) Classification of ore deposits. *In* Ore deposits of the western states. Am I M Eng, Pub 17:55, California 44-45 (1933) (*abst*) Rv Geol et Sci conn 14:566 (1934)

Bell, A. H.

31 Oil drainage and gas conservation problems in Kettleman Hills. (*ann*) An Bib Ec G 4:101 (1931)
32 (and Barnes, R. M.) Proration on the basis of uniform allowable gas production. *In* Petroleum development and technology. Am I M Eng, Tr 103:142-147 (1932)

Bennit, H. L.

31 (Mann, L.; Young, W. H.; Tryon, F. G.; and Berquist, F. E.) Coal. U S B M, Min Res, Calendar Year 1931:415-510, California 426, 472 (1933)
32 (Young, W. H.; Tryon, F. G.; and Mann, L.) Coal. U S B M, Min Yb, Stat App, Calendar Year 1932:373-454, California 376, 423 (1934)... 1933:281-360, California 286, 323 (1935)

Berkey, C. P.

31 (with Savage, J. L.; Louderback, G. D.; Hinderlider, M. C.; and Williams, I. A.) Report of consulting board on safety of the proposed Pine Canyon dam, Los Angeles County. Cal Dp Pub Works, Div Water Resources May 1931:22 pp, 8 pls, 1 map (1931)

Berquist, F. E.

31 (Young, W. H.; Tryon, F. G.; Mann, L.; and Bennit, H. L.) Coal. U S B M, Min Res, Calendar Year 1931:415-510, California 426, 472 (1933)

Berry, E. W.

29 Fossil *Meliosoma* from the Miocene of California. (*abst*) N Jb 1931, Ref 3:330 (1931)

Berthelot, Charles

34 Procedes de prospection, d'extraction et de traitement de minerais aurifères. Chimie et Industrie 31:3-31, 262-279 (1934) (*abst*) Rv Geol et Sci conn 14:351-352 (1934)

Beverly, Burt Jr.

34 Graphite deposits in Los Angeles County, California. Ec G 29:346-355 (1934) (*ann*) An Bib Ec G 7:68 (1935) (*abst*) Rv Geol et Sci conn 14:491 (1934) (*abst*) N Jb 1934, Ref 2:682 (1934) (*abst*) N Jb 1935, Ref 1:38 (1935) Chem Absts 28:6090-6091 (1934) (*abst*) G Zentralbl, Abt A, Bd 53:471 (1934)

Bignell, L. G. E.

33 Inglewood fault regarded as important factor in minimizing earthquake effect on oil fields. Oil Gas J 31:8, 29 (1933)

Billingsley, Paul

33 (and Locke, A.) Tectonic position of ore districts in the Rocky Mountain region. Am I M Eng, Tech Pub 501:12 pp (1933) (*ann*) An Bib Ec G 2:199-200 (1934) (*abst*) Eng Index 1933:729 (1933)
34 (Locke, A., and Schmitt, H. A.) Some ideas on the occurrence of ore in the western United States. Ec G 29:560-576 (1934)

Blackman, E. O.

35 Submarine geology from the air. Petroleum World An Rv 1935:187-188 (1935)

Blackwelder, Eliot

28 Notes on sedimentary deposits in the desert. (Rp committee on sedimentation) Nat Research Council, B 85:78-80 (1928) (*abst*) N Jb 1934, Ref 2:69 (1934)
29 The Kern River scarp. Am As G An 19:9-13 (1929)
31 Sandblast action in relation to the glaciers of Sierra Nevada. (*abst*) N Jb 1931, Ref 2:685 (1931)... Ref 3:1931:929-930 (1931)
31a Landslide family and its relations. (*abst*) G Soc Am B 42:296 (1931)
31b Specific evidence of deflation in deserts. (*abst*) N Jb 1931, Ref 2:348 (1931)
31c Pleistocene lakes of the Basin Range province. (*abst*) G Soc Am, B 42:313 (1931)
31d The lowering of playas by deflation. Am J Sc (5) 21:140-144 (1931) (*abst*) Zs Geom 7:64-65 (1932) (*abst*) N Jb 1934, Ref 2:68 (1934) Geog J 78:93-94 (1931) G Zentralbl 46:94-95 (1932) (*Review*) Geog J 78:93-94 (1931)

Blackwelder, Eliot—Cont.

31e Pleistocene glaciation in the Sierra Nevada and Basin Ranges. G Soc Am, B 42:865-922, 2 pls, 21 figs (1931) (*abst*) Zs Gletscherk 20:498-499 (1932) (*abst*) N Jb 1933, Ref 3:562-563 (1933)

32 Sedimentation studies at Stanford University. Nat Research Council, B 89: 99-100 (1932) (*abst*) N Jb 1934, Ref 2:69 (1934)

32a Origin of the Piedmont Plains of the Great Basin. (*abst*) G Zentralbl 46:93 (1932)

32b Glacial and associated stream deposits of the Sierra Nevada. Cal Dp Nat Res, Div Mines and Mining, Mining in California, St Mineralogist's Rp 28:303-310 (1932)

33 Eastern slope of the Sierra Nevada. Int G Cong, Guide Book 16:81-95 (1933) (*abst*) N Jb 1936, Ref 3:70-71 (1936)

33a Yardangs. (*abst*) Pan Am G 59:309 (1933) G Soc Am, B 45:159-165, 7 pls (1934) (*abst*) G Soc Am, Pr 1933:305 (1934)

34 Supplementary notes on Pleistocene glaciation in the Great Basin. Wash Ac Sc, J 24:217-222 (1934)

34a Talus slopes in Basin Range province. (*abst*) Pan Am G 61:313 (1934)

34b Origin of the Colorado River. G Soc Am, B 45:551-566, 3 figs (incl maps) (1934)

35 Pleistocene Lake Tecopa. (*abst*) Pan Am G 63:311 (1935) (*abst*) G Soc Am, Pr 1933:333 (1936)

36 (and **Ellsworth,** E. W.) Pleistocene lakes of the Afton Basin. Am J Sc 31 (5):453-463, figs (1936)

Blake, William Phipps

58 Report of a geological reconnaissance in California. N Y (1858)

67 Note upon partzite, Am J Sc (2) 44:119 (1867)

82 On the occurrence of vivianite in Los Angeles County. St Mineralogist's Rp 2:265 (1882)

Blanchard, F. B.

35 (and **Byerly,** P.) A study of a well gauge as a seismograph. Seism Soc Am, B 25:313-323 (1935)

35a (and **Byerly,** P.) Well gauges as seismographs. Nature 135:303-304 (1935)

Blaney, H. F.

30 (with **Taylor,** C. A. and **Young,** A. A.) Rainfall penetration and consumptive use of water in Santa Ana River Valley and Coastal Plain, California. Cal Dp Nat Res, Div Water Resources, B 3:158 pp (1930) (*ann*) An Bib Ec G 4:333 (1931)

Blank, E. W.

34 Diamond finds in the United States. Rocks and Min 9:147-150, 163-166, 179-182, California 148, 179-180 (1934)

Blount, A. L.

35 (**Hoots,** H. W. and **Jones,** P. H.) Marine oil shale, source of oil in Playa del Rey field, California. Am As Petroleum G, B 19:172-206 (1935) (*ann*) An Bib Ec G 8:142 (1936)

Blume, John A.

35 A machine for setting structures and ground into forced vibration. Seism Soc Am, B 25:361-381 (1935)

Boardman, Leona

32 (and **Mansfield,** G. R.) Nitrate deposits of the United States. U S G S, B 838: 107 pp. 11 pls, California 23-29 (1932) (*abst*) G Zentralbl, Abt A 48:413-414 (1932) Min Absts 5:164 (1932) (*ann*) An Bib Ec G 5:85 (1933)

Bode, Francis D.

31 Characters useful in determining the position of individual teeth in the permanent cheek-tooth series of *Merychippine* horses. (*abst*) Pan Am G 56:57 (1931) J Mammalogy 12:118-129, 13 figs (1931)

33 *Merychippine* species of western United States and their stratigraphic relationships. (*abst*) Pan Am G 59:377 (1933) G Soc. Am, Pr 1933 1:392 (1934) (*abst*) G Soc Am. Pr 1934:383 (1935)

33a (and **Findlay,** W. A.) Structures of a part of the San Joaquin Hills, California. (*abst*) Pan Am G 59:318-319 (1933) G Soc Am, Pr 1933:316 (1934)

34 *Merychippes* zone fauna, Coalinga, California. (*abst*) Pan Am G 62:68 (1934)

34a *Anchitheriine* horses from the *Merychippus* zone of the north Coalinga district, California. Carnegie Inst, Wash, Pub 440:43-58, 5 pls (1933) (*abst*) Pale Zentralbl 4:239 (1934) (*abst*) Pan Am G 58:149-150 (1932) (*abst*) G Soc Am, B 44: 219 (1933)

35 Tooth characters of *Protohippine* horses with special reference to species from the *Merychippus* zone, California. Carnegie Inst, Wash, Pub 453:39-63, 2 pls, 6 figs (1935)

35a The fauna of the *Merychippus* zone, north Coalinga district, California. Carnegie Inst, Wash, Pub 453:65-96, 2 pls, 10 figs (1935)

35b (and **Stock,** C.) Occurrence of lower Oligocene mammal-bearing beds near Death Valley, California. Nat Ac Sc, Pr 21:571-579, 2 pls (1935) (*abst*) G Zentralbl, Abt A, Bd 57:53-54 (1936)

Bodle, R. R.

29 (and **Hicks**, N. H.) United States earthquakes. U S Coast S 1928 Serial 483:31 pp, 5 figs (1930)... 1929 Serial 511:61 pp, California 11-15, 1 fig (1931)... 1930 Serial 539:25 pp, California 8-14, 2 figs (1932)... 1931 Serial 553:26 pp, California 14-20 (1932)... 1932 Serial 563:21 pp, California 6-15 (1934)

32 Earthquake notes. Seism Soc Am, Eastern section 3:12 pp (1932) (*abst*) Rv Geol et Sci conn 13:15 (1932)

32a (and **Newmann**, F.) United States earthquakes 1930, U S Coast S, Serial 539:25 pp (1932) (*abst*) Canada, Dom Obs, Pub, Bib Seism 10:274 (1932)

Bowers, N. A.

34 California makes progress against earthquake hazard. Eng News Rec 113: 14-17 (1934) (*abst*) Eng Index 1934:323 (1934)

Bowes, Glenn H.

30 (and **Barnes, R. M.**) Seal Beach oil field. Cal Dp Nat Res, Div Oil and Gas, California Oil Fields 16, no 2:9-31, 9 figs, 8 pls (1930) (*abst*) Rv Geol et Sci conn 13:234 (1933) (*abst*) Eng Index 1932:899 (1932)

Bowie, A. J.

93 A practical treatise on hydraulic mining in California. (Fifth edition) 313 pp, 72 figs (1893)

Bowie, William

35 Fundamental geodetic surveys in the United States nearing completion. Nat Ac Sc, Pr 21:32-36, California 34, 35 (1935)

36 Vertical movements of earth's crust as determined by leveling. J G 44:387-395 (1936) (*abst*) Rv Geol et Sci conn 16:347-348 (1936)

Bowles, Oliver

31 Chalk, whiting and whiting substitutes, U S B M, Inf Circ 6482:13 pp (1931) (*abst*) U S B M, List Pub 1910-1932:149 (1933)

31a (and **Hatmaker**, P.) Trends in the production and uses of granite as dimension stone. U S B M, Rp Invest 3065:21 pp, 11 figs (1931) (*abst*) U S B M, List Pub 1910-1932:115 (1933)

31b Talc and soapstone. Min Ind 39:572-578, California 572, 575 (1931)... 40:520-525, California 522 (1932)... 41:497-502, California 498 (1933)

32 Abrasive materials. U S B M, Min Res, Calendar Year 1929:65-81, California 66, 74, 75, 76, 77 (1932)

Bowles, Oliver—Cont.

32a (and **Coons, A. T.**) Slate. U S B M, Min Res, Calendar Year 1929:161-174, California 166 (1932)... 1930:277-290, California 283, 286, 290 (1932)... 1931:165-177, California 171, 173, 177 (1933) (Stat App) Min Yb 1932-1933:13-16, California 14 (1934) Min Yb 1934:817-828, California 822, 826, 827 (1935)... 1936:833-840, California 837, 839 (1934)

32b (and **Middleton**, J.) Feldspar. U S B M, Min Res, Calendar Year 1929:83-93, California 89 (1932)

32c (and **Stoddard**, B. H.) Asbestos. U S B M, Min Res, Calendar Year 1929: 195-207, California 199 (1932)... 1930:263-275, California 266, 267 (1932)... Min Yb 1934:1009-1016, California 1012 (1934)

32d (and **Stoddard**, B. H.) Talc and soapstone. U S B M, Min Res, Calendar Year 1930:303-313, California 304, 305, 306, 307, 308 (1932)... 1931:99-110, California 100, 103, 104, 110 (1933)

32e Asbestos. Min Ind 40:48-56, California 52 (1932)

33 (and **Banks**, D. M.) Onyx marble and travertine. U S B M, Inf Circ 6751: 1-12, California 5, 9 (1933)

33a (and **Justice**, C. W.) Growth and development of the nonmetallic mineral industries. U S B M, Inf Circ 6687:1-50, California 10-12, 29-36, 40-42, 44-46, 49 (1933)

34 Asbestos—domestic and foreign deposits. U S B M, Inf Circ 6790:1-24, California 3, 5 (1933)

34a (and **Davis**, A. E.) Abrasive materials. U S B M, Min Yb 1934:889-906, California 891, 892 (1934)

34b The stone industries. 519 pp, tables, figs, maps (1934) Crushed stone industry 400, 401, 478, 479, 480. Basalt 477, 478. Sandstone 73, 476, 478, 479. Limestone 399, 400, 401, 203. Diatomite 344. Granite 137, 475, 478. Marble 202, 203, 401. Lapis-lazuli 345. Travertine 44, 401. Felsite porphyry 479. Soapstone 291. Onyx 203. Slate 251. Cement 399, 400.

35 (and **Coons**, A. T.) Lime. U S B M, Min Yb 1935:967-976, California 969 (1935)

36 (and **Banks**, D. M.) Lime. U S B M, Inf Circ 6884:37 pp California 11, 21, 24 (1936)

Bradley, James

31 Mining and milling at the Spanish Mine. M Metal 12:435-439 (1931) Chem Absts 26:53 (1932)

Bradley, Walter W.

30 Biennial report of the State Mineralogist. Cal Dp Nat Res, Div Mines,

Bradley, Walter W.—Cont.

Mining in California, St Mineralogist's Rp 26:489-494 (1930)

30a Barite in California. Am I M Eng, Tr 1931:170-176 (1931) (ann) An Bib Ec G 4:272 (1931) (abst) M Metal 11:18 (1930) Cal Dp Nat Res, Div Mines, Mining in California, St Mineralogist's Rp 26:45-57 (1930)

32 California geological surveys. In summary information on the State Geological Surveys and the United States Geological Survey under the direction of M. M. Leighton. Nat Res Council, B 88: 10-14 (1932)

32a Renewed activity in California gold mining. M Metal 13:385-390 (1932) (abst) Am I Eng, Tr 102:266-267 (1932) (abst) Eng Index 1932:634 (1932)

32b Sanbornite, a newly described mineral from California. Cal Dp Nat Res, Div Mines and Mining, Mining in California 28:82-83 (1932)

32c Biennial report of State Mineralogist. Cal Dp Nat Res, Div Mines, Mining in California, St Mineralogist's Rp 28:385-394 (1932)

33 Itinerary, Yosemite to Mother Lode. Int G Cong, Guide Book 16:62-63 (1933)

34 An echo of the days of '49. Eng M J 135:494-496 (1934) (ann) An Bib Ec G 7: 243 (1934)

34a The nonmetallic minerals of California. Pit and Quarry 26 (11):35-40 (1934) (ann) An Bib Ec G 7:56 (1935)

34b Biennial report of the State Mineralogist, Cal Dp Nat Res, Div Mines, M G, J, St Mineralogist's Rp 30:431-439 (1934)

35 Change made in California law for mining claims. M Metal 16:389 (1935)

35a Recent nonmetallic mineral development in California. M Metal 16:181-184 (1935) (ann) An Bib Ec G 8:75 (1935) (abst) N Jb 1935, Ref 1:555 (1935) (abst) Am I M Eng, Yb 1935:58 (1936)

35b California gold mining second to oil in value for year 1934. Cal Oil World 28, no 16:76 (1935)

36 Biennial report of the State Mineralogist, Cal Dp Nat Res, Div Mines, M G, J, St Mineralogist's Rp 32:481-489 (1936)

Bradley, Worthen W.

31 Method and cost of recovering quicksilver from low-grade ore at the reduction plant of the Sulphur Bank Syndicate, Clear Lake, California. U S B M, Inf Circ 6429:17 pp, 11 figs (1931)

Bramkamp, R. A.

34 Molluscan fauna of Imperial formation of San Gorgonio Pass. (abst) Pan Am G 62:70-71 (1934) (abst) G Soc Am, Pr 1934:385 (1935)

Bramlette, M. N.

33 Rhythmic bedding in the Monterey rocks of California. (abst) Wash Ac Sc, J 23:575 (1933)

33a Petrographic studies in correlation of sands from Kettleman Hills, California. (abst) Pan Am G 59:235 (1933)

33b (and Ponsjak, E.) Zeolitic alteration of pyroclastics. Am Mineralogist 18:167-171, 1 fig (1933) Min Absts 5:357 (1933)

34 Heavy mineral studies on correlation of sands at Kettleman Hills, California. Am As Petroleum G, B 18:1559-1576 (1934) (abst) N Jb 1935, Ref 2:696 (1935) (abst) Rv Geol et Sci conn 15:7 (1935)

36 (and Woodring, W. P., Kleinpell, R. M.) Miocene stratigraphy and paleotology of the Palos Verdes Hills, California. Am As Petroleum G, B 20:125-149 (1936) (abst) G Zentralbl Abt A, Bd 57:51-52 (1936) (abst) Rv Geol et Sci conn 16:378 (1936)

Bratter, Herbert M.

36 Silver. Eng M J 137:55-56, 62 (1936) •

Breakey, H. A.

36 (White, A. G., and Hopkins, G. R.) Crude petroleum and petroleum products. U S B M, Min Yb 1936:667-723, California 668, 674, 678, 681, 686, 687, 691, 693, 695, 711 713 (1936)

Bremner, Carl St. J.

32 Geology of Santa Cruz Island, Santa Barbara County, California. Santa Barbara Mus N H, Oc P 1:33 pp, 12 figs, 5 pls (1932)

33 Geology of San Miguel Island, Santa Barbara County, California. Santa Barbara Mus N H, Oc P 2:23 pp, 10 figs, 4 pls (incl map) (1933)

Bridge, A. F.

36 Dry natural-gas reserves, their control and conservation, a California problem. M Metal 17:393-395 (1936)

Brooks, Benjamin T.

36 Origins of petroleums: chemical and geochemical aspects. Am As Petroleum B, B 20:280-300, California 281, 287-288, 289, 292, 293 (1936)

36a (and Snider, L. C.) Probable petroleum shortage in the United States, and methods for its alleviation. Am As Petroleum G, B 20:15-50, California 24, 27-28, 34, 36, 40 (1936)

Brown, Arthur B.

32 (and Kew, W. S. W.) Occurrence of oil in metamorphic rocks of San Gabriel Mountains, Los Angeles County, California. Am As Petroleum G, B 16:777-785,

Brown, Arthur B.—Cont.

4 figs (1932) (*abst*) Rv Geol et Sci conn 13:237 (1933) G Zentralbl, Abt. A 48:317 (1932) Inst Pet Tech, J 18:415A (1932) Chem Absts 26:5278 (1932)

Brown, C. C.

31 Kettleman Hills most important California gas reserve. Oil Weekly 63, no 1:28... 2:26, 28 (1931) (*ann*) An Bib Ec G 4:317 (1931)

Brown, J. S.

27 Routes to desert watering places in the Salton Sea region, California. (*abst*) N Jb 1927, Ref 2, Abt B:113 (1927)

Brown, S. M.

33 (**Kelley**, W. P.; and **Woodford**, A. C.) Clay minerals of California soils. (*abst*) Pan Am G 59:315-316 (1933)

Brues, C. J.

28 Studies on the fauna of hot springs in the western United States and the biology of *Thermophilous* animals. Am Ac Arts, Pr 63:139-228, pls, figs (1928)

Bryan, E. N.

32 Topographic mapping program for California. Cal Dp Nat Res, Div Mines, Mining in California, St Mineralogist's Rp 28:85-87 (1932)

Bryan, Kirk

31 Physiographical study in the Salton Basin Region. Geog Rv 21:153-154 (1931) (*abst*) G Zentralbl 44:496 (1931)

31a Wind-worn stones or ventifacts—a discussion and bibliography. Rp Committee Sedimentation for 1929-1930 Nat Res Council Rp and Circ Sed no 98:29-50 (1931) (*abst*) Zs Geom 8:144 (1934)

31b (and **Wickson**, G. G.) The W. Penck method of analysis in southern California. Zs Geom 6:287-291 (1931)

Buddington, A. F.

33 Correlation of kinds of igneous rocks with kinds of mineralization. *In* ore deposits of the western states. Am I M Eng, Pub 3:350-385 (1933)

Bue, C. D.

34 (**Pritchett**, H. C.; and **Piper**, A. M.) Seepage loss and gain of the Mokelumne River, California, U S G S Memorandum for the Press (mimeographed) (Ph 85246) (1934)

Burchard, E. F.

33 Fluorspar deposits in western United States. Am I M Eng, Tech Pub 500:26 pp, map (1933)

Burchfiel, B. M.

36 Ceramic materials other than clays abundant in California. M Metal 17:441-443 (1936)

Burgess, J. A.

35 Mining gold on Carson Hill. Eng M J 136:111-115 (1935) (*ann*) An Bib Ec G 8:56 (1935) (*abst*) G Zentralbl, Abt A, Bd 57:157 (1936)

Bush, Edgar

32 Active placers in southern California. M J Ariz 16, no 5:7 (1932)

Bush, J. Burchard

30 Foraminifera of Tomales Bay, California. Stanford Univ, Micropaleontology B 2:38-43 (1930)

Bush, R. D.

30 Fifteenth annual report of the State Oil and Gas Supervisor of California for the fiscal year 1929-1930, Cal Dp Nat Res, Div Oil and Gas, California Oil Fields... Sixteenth 1930-1931... Seventeenth 1931-1932... Eighteenth 1932-1933... Nineteenth 1933-1934... Twentieth 1934-1935

32 Resume of oil field operations in 1931. Cal Dp Nat Res, Div Oil and Gas, California Oil Fields 17:164 pp (1932) (*abst*) Eng Index 1933:774 (1933)

Butler, B. S.

32 Influence of the replacement minerals associated with ore deposits. Ec G 27:1-24 (1932)

33 Ore deposits as related to stratigraphic structural and igneous geology in the western United States. *In* ore deposits of the western states. Am I M Eng, Pub:198-240 (1933)

33a Ore deposits of the United States in their relation to geologic cycles. Ec G 28:301-328 (1933)

Buttgenback, H.

32 Notes mineralogiques. Soc G Belgique, An 55:B165-B178, 6 figs (1932)

Butts, Allison

35 Lead. Min Ind 43:362-389, California 366-367 (1935)

Buwalda, John Peter

31 Reversal in direction of vertical component movement along faults. (*abst*) Pan Am G 55:64 (1931)

31a (and **Wood**, H. O.) Horizontal displacement along the San Andreas fault in the Carrizo Plain, California. (*abst*) G Soc Am, B 42:298-299 (1931) Pan Am G 54:75 (1930)

Buwalda, John Peter—Cont.

31b (**Gazin,** C. L.; and **Sutherland,** J. C.) Frazier Mountain; a crystalline overthrust slab without roots, west of Tejon Pass, southern California. (*abst*) G Soc Am, B 42:294-295 (1931) (*abst*) N Jb 1933, Ref 2:251 (1933)

32 (**Gutenberg,** B.; and **Wood,** H. O.) Experiments testing seismographic methods for determining crustal structure. (*abst*) Pan Am G 58:65-66 (1932) Seism Soc Am, B 22:185-242 (1932)

33 The Long Beach earthquake. Science n s 78:148-149 (1933)

34 Tertiary tectonic activity in the Tehachapi region. (*abst*) Pan Am G 61: 309-310 (1934) (*abst*) G Soc Am, Pr 1934: 312 (1935)

35 (and **Gutenberg,** B.) Seismic reflection profile across Los Angeles basin. (*abst*) Pan Am G 63:303 (1935) (*abst*) G Soc Am, Pr 1935:327 (1936)

35a (and **Gutenberg,** B.) Investigation of overthrust faults by seismic methods. Science, n s 81, no 2103:384-386 (1935)

Byerly, Perry

27 Registration of earthquakes at the Berkeley Station and at the Lick Observatory Station from April 1, 1925 to September 30, 1925... October 1, 1925 to March 31, 1926. Cal Univ, Seism Sta, B 2:156-199 (1927)

27a (and **Jones,** A. E.) Registration of earthquakes at the Berkeley Station and at the Lick Observatory Station from April 1, 1926 to September 30, 1926... October 1, 1926 to March 31, 1927. Cal Univ, Seism Sta, B 2:202-250 (1927)

28 The registration of earthquakes at the Berkeley Station and at the Lick Observatory from April 1, 1927 to September 30, 1927. Cal Univ, Seism Sta, B 2:251-272 (1928)

28a (and **Dyk,** R.) Registration of earthquakes at the Berkeley Station and at the Lick Observatory Station from October 1, 1927 to March 31, 1928. Cal Univ, Seism Sta, B 2:273-300 (1928)

30 (and **Dyk,** R.) Registration of earthquakes at the Berkeley Station and at the Lick Observatory Station from April 1, 1929 to September 30, 1929. Cal Univ, Seism Sta, B 2:361-397 (1930)

30a (and **Dyk,** R.) Registration of earthquakes at the Berkeley Station and at the Lick observatory Station from October 1, 1929 to March 31, 1930. Cal Univ, Seism Sta, B 2:399-439 (1930)

30b The California earthquake of November 4, 1927. Seism Soc Am, B 20:54-66 (1930)

31 The earthquakes of November 28, 1929, and the surface layers of the earth

Byerly, Perry—Cont.

in California. Nat Ac Sc, Pr 17:91-100, 2 figs (1931) (*abst*) Canada, Dom Obs, Pub, Bib Seism 10:160 (1931) (*abst*) Science n s 72:373 (1930) Nature 127:995 (1931)

31a (**Hester,** J., and **Marshall,** K.) The natural periods of vibration of some tall buildings in San Francisco. Seism Soc Am, B 21:268-276, 3 figs (1931) (*abst*) Canada, Dom Obs, Pub, Bib Seism 10:228 (1932)

31b Dispersion of seismic waves of the Love type and the thickness of the surface layer of the earth under the Pacific. G Soc Am, B 42:312 (1931) Beitr Geoph 26:27-33 (1930)

31c The registration of earthquakes at the Berkeley station and at the Lick Observatory station from April 1, 1930 to September 20, 1930. Cal Univ Seism Sta, B 2:442-476 (1931)... October 1, 1930 to March 31, 1931, B 2:477-533 (1932)... April 1, 1931 to September 20, 1931, B 3:1-51 (1932)

31d Seismology at the University of California. Seism Soc Am, Eastern Sec, Earthquake Notes 3:9-10 (1931)

32 (and **Sparks,** N. R.) Earthquakes in northern California and the registration of earthquakes at Berkeley, Mount Hamilton, Palo Alto from October 1, 1931 to March 31, 1932. Cal Univ, Seism Sta B 3:53-96 (1933)... April 1, 1932 to September 30, 1932, B 3:97-150 (1933)... October 1, 1932 to March 31, 1933, B 3:151-241 (1935)

32a (and **Dyk,** K.) Richmond quarry blast of September 12, 1931 and the surface layering of the earth in the region of Berkeley. Seism Soc Am, B 222:50-55 (1932)

33 Seismic geography. Nat Res Council, B 90:206-215 (1933)

33a (and **Sparks,** N. R.) The first preliminary waves of the California earthquake of June 6, 1932. Seism Soc Am, Eastern Sec, Earthquake Notes 5:254-256, fig (1933) Am Geop Union, Tr 14:254-256 (1933)

33b California earthquakes. (*abst*) Am Soc Civil Eng. 169th Regular Meeting of the San Francisco Section, Pr 1933:2-3 (1933)

33c Vibrations of buildings. Am Soc Mechanical Eng, Pacific Coast Applied Mechanics Meeting, Pr 1933:16 (1933)

33d (and **Sparks,** N. R.) Earthquakes in northern California and the registration of earthquakes at Berkeley, Mt. Hamilton, Palo Alto from October 1, 1931 to March 31, 1932. Cal Univ, Seism Sta, B 3:54-96 (1933)

33e (and **Sparks,** N. R.) Earthquakes in northern California and the registration of earthquakes at Berkeley, Mt.

Byerly, Perry—Cont.

Hamilton, Palo Alto, San Francisco from April 1, 1932 to September 30, 1932. Cal Univ, Seism Sta, B 3:97-150 (1933)

34 Northern California earthquakes, April 1, 1932 to April 1, 1933. Seism Soc Am, B 24:115-117 (1934)

35 (and Wilson, J. T.) Earthquakes in northern California and the registration of earthquakes at Berkeley, Mount Hamilton, Palo Alto, and San Francisco from April 1, 1933 to September 30, 1933. Cal Univ Seism Sta, B 4:1-73 (1935)

35a (and Wilson, J. T.) Earthquakes in northern California and the registration of earthquakes at Berkeley, Mount Hamilton, Palo Alto, San Francisco, Ferndale from October 1, 1933 to March 31, 1934. Cal Univ Sta, B 4:75-165 (1935)... April 1, 1934 to September 30, 1934, B 4: 167-243 (1936)... October 1, 1934 to March 31, 1935, B 4:245-338 (1936)

35b (and Wilson, J. T.) The Richmond quarry blast of August 16, 1934. Seism Soc Am, B 25:259-268 (1935)

35c (and Blanchard, F. B.) A study of a well gauge as a seismograph. Seism Soc Am, B 25:313-321 (1935)

35d (and Sparks, N. R.) Earthquakes in northern California and the registration of earthquakes at Berkeley, Mt. Hamilton, Palo Alto, San Francisco from October 1, 1932 to March 31, 1933. Cal Univ, Seism Sta, B 3:151-241 (1935)

35e (and Wilson, J. T.) Northern California earthquakes, April 1, 1933 to March 31, 1934. Seism Soc Am, B 25:269-273 (1935)

35f (and Blanchard, F. B.) Well gauges as seismographs. Nature 135:303-304 (1935)

35g (and Wilson, J. T.) The central California earthquakes of May 16, 1933 and June 7, 1934. Seism Soc Am, B 25: 223-347 (1935)

36 (and Annis, W.) Earthquakes in northern California and the registration of earthquakes at Berkeley, Mount Hamilton, Palo Alto, San Francisco, Ferndale from July 1, 1935 to September 30, 1935. Cal Univ, Seism Sta, B 5:40-78 (1936)... October 1, 1935 to December 31, 1935, B 5:80-117 (1936)

36a (and Hoskins, E. E.) Earthquakes in northern California and the registration of earthquakes at Berkeley, Mount Hamilton, Palo Alto, San Francisco, Ferndale From January 1, 1936 to March 31, 1936. Cal Univ, Seism Sta, B 6:1-37 (1936)... April 1, 1936 to June 30, 1936, B 6:38-85 (1936)

California Department of Public Works

31 Kings River Water Master Report for period 1918-1930. Cal Dp Pub Works,

California Department of Public Works— Cont.

Div Water Resources, B 38:426 pp, pl (1931) (abst) Eng Index 933:1219 (1933)

31a Variation and control of salinity in Sacramento-San Joaquin Delta and upper San Francisco Bay. Cal Dp Pub Works, Div Water Resources, B 27:440 pp, illus, charts, tables, maps (1932)

32 Records of ground water levels at wells, South Coastal Basin investigation, 1932. Cal Dp Pub Works, Div Water Resources B 29:590 pp, maps, tables (1932) (abst) Eng Index 1933:1221 (1933)

32a Economic aspects of a salt water barrier below confluence of Sacramento and San Joaquin Rivers. Cal Dp Pub Works, Div Water Resources, B 28:450 pp, illus, charts, tables, maps (1932)

33 Sacramento River basin, 1931. Cal Dp Pub Works, Div Water Resources, B 26:583 pp, illus, charts, tables, maps (1933) (abst) Eng Index 1933:1919 (1933)

33a Quality of irrigation waters, South Coastal Basin investigation, 1933. Cal Dp Pub Works, Div Water Resources, B 40:95 pp, tables, maps (1933)

33b Pit River investigation, 1933. Cal Dp Pub Works, Div Water Resources, B 41:152 pp, tables, charts, maps (1933)

33c Records of ground water levels at wells for the year 1932. Seasonal precipitation records to and including 1931-1932. Cal Dp Pub Works, Div Water Resources, B 39A:162 pp, map (mimeographed) (1933)

34 San Joaquin River basin, 1931. Cal Dp Pub Works, Div Water Resources, B 29:656 pp, illus, charts, tables, maps (1934)

34a Ventura County investigation, 1933. Cal Dp Pub Works, Div Water Resources, B 46:244 pp, tables, charts, maps (1934) (abst) An Bib Ec G 8:162 (1936) (abst) G Zentralbl, Abt A, Bd 57:501 (1936)

34b Ventura County investigation. Basic data for the period 1927-1932, incl. Cal Dp Pub Works, Div Water Resources, B 46A (mimeographed) 574 pp, tables, map (1934) (abst) An Bib Ec G 8:162 (1936) (abst) G Zentralbl, Abt A, Bd 57: 501 (1936)

34c Mojave River investigation, 1934. Cal Dp Pub Works, Div Water Resources, B 47:247 pp (mimeographed), tables, charts, maps (1934)

34d San Joaquin River basin, 1931. Cal Dp Pub Works, Div Water Resources, B 29:656 pp, illus, charts, tables, maps (1934)

34e Records of ground water levels at wells for the year 1933. Seasonal precipitation records to and including 1932-1933. Cal Dp Pub Works, Div Water Resources, B 39B:146 pp (mimeographed) (1934)

California Department of Public Works— Cont.

34f Water losses under natural conditions from wet areas in southern California. South Coastal Basin investigation, 1933. Cal Dp Pub Works, Div Water Resources, B 44:176 pp, illus, charts, tables, maps (1934)

35 San Diego County investigation, 1935. Cal Dp Pub Works, Div Water Resources, B 48:300 pp (mimeographed) tables, charts, maps (1935)

35a Records of ground water levels at wells for the year 1934. Seasonal precipitation records to and including 1934-1935. Cal Dp Pub Works, Div Water Resources, B 39C:148 pp (mimeographed) (1935)

35b Geology and ground water storage capacity of valley fill. South Coastal Basin investigation, 1934. Cal Dp Pub Works, Div Water Resources, B 45:280 pp, illus, charts, tables, maps (1935)

36 Records of ground water levels at wells for the year 1935. Cal Dp Pub Works, Div Water Resources, B 39D (mimeographed) 122 pp (1936)

California Oil World

31 Sespe producer aids geologist. Cal Oil World 24, no 13:1, 5 (1931)

33 Stalder says Buttes drilling violates text book geology but fits experience. Cal Oil World 25:12, 33 (1933)

33a Three groups seek carbon dioxide gas east of Salton Sea in Imperial Valley. Cal Oil World 25:15, 30 (1933)

33b El Monte wildcat will test interesting geological theories. Cal Oil World 25, no 41:11 (1933)

33c Huge deposits of diatomaceous earth may supply billion barrels of oil. Cal Oil World 26, no 22:4 (1933)

34 Temblor found at 750 feet in San Luis Obispo County well. Cal Oil World 26, no 51:5 (1934)

34a New geophysical methods claimed to determine oil, geology. Cal Oil World 27, no 12:11 (1934)

35 Geological and legal history of Del Rey area outlined. Cal Oil World 27, no 29:15 (1935)

35a Unsung source of oil in diatomaceous earth; deposit at Casmalia. Cal Oil World 28, no 16:24 (1935)

Calkins, F. C.

32 The granitic rocks of the Yosemite region. (abst) G Zentralbl 46:401-402 (1932)

Callaghan, Eugene

34 (and **Gianella**, V. P) The earthquake of December 20, 1932 at Cedar Mountain, Nevada, and its bearing on the genesis of Basin range structure. J G 42:1-22

Callaghan, Eugene—Cont.

(1934) (Review) Sc Progress 29:501 (1935) (abst) G Zentralbl 51:455 (1934)

Calvert, Earl L.

33 Boron in California. Miner Soc S Calif, B 2:1-2 (1933)

34 Kernite supplies the world with borax. Oregon Mineralogist 2:18-19 (1934)

Cameron, E. N.

36 (and **Kerr**, P. F.) Fuller's earth of bentonitic origin from Tehachapi, California. Am Mineralogist 21, no 4:230-237 (1936)

Campbell, A. B.

34 (and **Allen**, R. C.) Story of Grass Valley—past and present. M J, Ariz 17, no 18:3-4, 24 (1934) (abst) Eng Index 1934: 536 (1934)

Campbell, Arthur S.

34 (and **Cushman**, J. A.) A new *Spiroplectoides* from the Cretaceous of California, Cushman Lab Foram Res, Contr 10:70-71 (1934)

35 (and **Cushman**, J. A.) Cretaceous foraminifera from the Moreno shale of California. Cushman Lab Foram Res, Contr 11:65-73, figs (1935) (abst) Rv Geol et Sci conn 15:606 (1936)

Campbell, C. D.

35 (and **Waters**, A. C.) Mylonites from San Andreas fault zone. (abst) G Soc Am, Pr 1934:325 (1935) (abst) Pan Am G 61:319-320 (1934) Am J Sc (5) 29:473-503 (1935) (abst) Rv Geol et Sci conn 15:359-360 (1935) (abst) Sc Progress 30:307 (1936) Chem Absts 29:6182 (1935)

Canfield, N. C.

31 (**Grover**, N. C., **McGlashan**, H. D., and **Henshaw**, F. F.) Surface water supply of the United States, 1928 Part XI Pacific Slope basins in California. U S G S, W-S P 671: IX, 304 pp (1931)... 1929, 691:IX, 294 pp (1931)... 1930, 706:IX, 317 pp (1932)... 1931, 721:XI, 497 pp (1932)... 1932, 736:XI, 415 pp (1933)

Carder, Dean S.

34 Seismic surface waves and the crustal structure of the Pacific region. Seism Soc Am, B 24:231-302 (1934)

Carlson, Anders Johan

31 Geothermal variations in oil fields of Los Angeles Basin. California. (abst) G Zentralbl 43:238 (1930-1931)

31a Geothermal variations in Coalinga area, Fresno County, California. Am As Petroleum G, B 15:829-836, 3 figs (1931)

Carlson, Anders Johan—Cont.

(abst) Rv Geol et Sci conn 12:255 (1932)...
13:100 (1933) (ann) An Bib Ec G 4:301
(1931) (abst) G Zentralbl 45:355 (1931)
Inst Pet Tech, J 17:425A-426A (1931)
(abst) Eng Index 1931:1036 (1931) Zs Prak
G 40:94 (1932) (abst) N Jb 1932 Band 2
Ref 2:537 (1932) (abst) U S B M, Geophysical Absts, Inf Circ 6568:256 (1931)

Carnegie Institute, Washington

32 The petrified forest of California.
Carnegie Inst Wash, News Service B,
School Ed 11, no 14:102-104 (1932)

Cassell, Dorothy

33 (and **Tieje,** A. J.) Megafauna and
micro-fauna of the Pleistocene and Pliocene formations of southern California as
revealed in a deep well near Ventura.
(abst) Pan Am G 59:376 (1933) G Soc
Am, Pr 1933: 1:390-391 (1934)

Cattell, R. A.

35 (and **Fowler,** H. C.) Influence of
technology upon the composite interest in
oil. U S B M, Min Yb 1935:771-794, California 773, 778, 789-791, 793-794 (1935)

Chalfant, W. A.

31 Death Valley. (Review) Nature 127:
368 (1931)

Chamberlin, Rollin T.

31 Isostasy from the geological point of
view. J G 39:1-23 (1931) (abst) Canada,
Dom Obs, Pub, Bib Seism 10:160-161
(1931)

Chandler, J. W.

34 Lava cap enterprise newest gold
producer in Nevada City-Grass Valley
district. Eng M J 135:362-365 (1934)
(abst) Eng Index 1934:536 (1934) (abst)
G Zentralbl, Abt A, Bd 55:457 (1935)

Chaney, Ralph W.

31 Age of the auriferous gravels.
(abst) G Soc Am, B 42:289-320 (1931)
43:226-227 (1932) (abst) Pan Am G 55:361
(1931)... 56:71-72 (1931) (ann) An Bib Ec
G 5:277 (1932)
32 Notes on occurrence and age of
fossil plants found in the auriferous
gravels of Sierra Nevada. Cal Dp Nat
Res, Div Mines and Mining, Mining in
California, St Mineralogist's Rp 28:299-302
(1932)
33 Further evidence regarding the age
of the auriferous gravels. (abst) G Soc
Am, B 44:78 (1933)
33a (and **Dorf,** E.) Ecology of the Tertiary forests of western North America.
(abst) G Soc Am, Pr 1933 1:357 (1934)

Chaney, Ralph W.—Cont.

33b (and **Mason,** H. L.) A Pleistocene
flora from the asphalt deposits at Carpinteria, California. Carnegie Inst,
Wash, Pub 415:45-79, 9 pls (1933) Biol
Absts 9:434, entry 3721 (1935) (abst) Pale
Zentralbl 4:314 (1934)
34 (and **Mason,** H. L.) A Pleistocene
flora from Santa Cruz Island, California.
Carnegie Inst, Wash, Pub 415:1-24, 7
pls, 1 fig (1934) (abst) N Jb 1931, Ref
3:869 (1931) (abst) G Zentralbl, Abt A
48:32 (1932) (abst) Pale Zentralbl 1:313
(1932)... 2:143-144 (1932)
35 Age of Clarno formation. (abst)
Pan Am G 64:71 (1935)
36 (and **Clements,** F. E.) Environment and life in the Great Plains. Carnegie Inst, Wash, Supplementary Pub 24:
54 pp, California 16, 17, 18-19, 22, 25, 27,
42, 46, 48, figs, pls (1936)

Chang, G. L.

31 The active and recently extinct volcanoes of North America. The Volcano
Letter, no 363:1-2 December 10 (1931)

Chapman, E. W.

32 Marine fossils in Imperial County,
California. Rocks & Min 7:53 (1932)
32a The Devil's Post Pile. Rocks &
Min 7:134 (1932)

Chappius, L. C.

31 Highlights of Los Alamos field.
Cal Oil World 23:10 (1931)

Chelikowsky, J. R.

36 (**Mayo,** E. B.; and **Conant,** L. C.)
Southern extension of the Mono Craters,
California. Am J Sc 32 (5):81-97, figs,
maps (1936)

Chellson, H. C.

34 What will it cost to work gold
placer of medium size? Eng M J 135:
441-445 (1934) (abst) Eng Index 1934:540
(1934)

Chick, C. A.

33 Long Beach earthquake of March 10,
1933 and its effect on industrial structures. Am Geop Union, Tr 14:273-284
(1933) Seism Soc Am, 5:273-284, 22 figs
(1933)

Church, C. C.

31 (and **Cushman,** J. A.) Some upper
Cretaceous foraminifera from near Coalinga. (abst) N Jb 1931, Ref 3:76-77
(1931)
31a The occurrence of *Kyphopyxa* in
California. (abst) G Zentralbl 43:330
(1930-1931)

Church, C. C.—Cont.

31b Cretaceous-Eocene contact north of Coalinga, California. Am As Petroleum G, B 15:697-699, fig (1931) (abst) G Zentralbl 45:267 (1931) (abst) N Jb 1932, Ref 2:113 (1932)

31c Foraminifera of the Lillis shale. (abst) G Soc Am, B 42:305-306 (1931) (abst) Pale Zentralbl 1:26 (1932)

31d Foraminifera of the Kreyenhagen shale. Cal Dp Nat Res Div Mines and Mining, Mining in California, St Mineralogist's Rp 27:202-213, 3 pls (1931) (abst) Rv Geol et Sci conn 13: 527-528 (1933) Pale Zentralbl 1:25 (1932)

Clark, Alex

31 The cool-water Timms Point Pleistocene horizon at San Pedro, California. San Diego Soc N H, Tr 7:25-42 (1931) (abst) Rv Geol et Sci conn 12:594 (1932) (abst) Pale Zentralbl 3:104 (1933)

33 Environment of marine mollusca living off Long Beach, California, and its bearing on Pleistocene correlations. (abst) Pan Am G 59:379 (1933) (abst) G Soc Am, Pr 1933 1:393-394 (1934)

35 (and **Clark, L. M.**) The Vaqueros in the Temblor Range. (abst) Am As Petroleum G, B 19:137 (1935)

Clark, Bruce Lawrence

28 The stratigraphy and faunal relationships of the Meganos group, middle Eocene of California. (abst) N Jb 1928, Ref 1, Abt A:362 (1928)

31 Classification of physiographic surfaces. (abst) Pan Am G 55:366-367 (1931)

31a Questioned boundaries for the marine Oligocene of western North America. (abst) G Soc Am, B 43:289-290 (1932) (abst) Pan Am G 56:69-70 (1931)

31b Fault through sedimentation. (abst) G Soc Am, B 43:230 (1932) (abst) Pan Am G 55:371-372 (1931)

31c Stratigraphic relationships in the Mount Diablo area of the upper Eocene deposits to those of the Oligocene. (abst) G Soc Am, B 42:304 (1931) (abst) Pan Am G 54:78 (1930)

31d Position of the fauna of the Astrodapsis antiselli zone. (abst) G Soc Am, B 43:289-290 (1932) (abst) Pan Am G 56: 68-69 (1931)

32 Classification of physiographic types in the Coast Ranges of California. (abst) G Soc Am, B 43:230-231 (1932)

32a Pliocene sequence in Berkeley Hills. (abst) Pan Am G 59:379 (1933) (abst) G Soc Am, B 44:151-152 (1933)

32b A new family and new genus from the Tertiary of the Pacific Coast. (abst) N Jb 1931, Ref 3:962-963 (1932)

Clark, Bruce Lawrence—Cont.

32c Tectonics of the Coast Ranges of middle California. (abst) G Zentralbl 46: 244 (1932) (abst) Eng Index 1931:655 (1931)

33 The Berkeley hills. Int G Cong, Guide Book 16:21-26 (1933) (abst) N Jb 1936, Ref 3:66-67 (1936)

33a A new genus and two new species of lamellibranchiata from the middle Eocene of California. J Paleontology 8: 270-272 (1934) (abst) G Soc Am, Pr 1933: 377 (1934) (abst) N Jb 1935, Ref 3:258-259 (1935) (abst) Rv Geol et Sci conn 15: 424-425 (1935)

34 Folding by drag and fault through deposition in Mt. Diablo and Coalinga areas of Coast Ranges. (abst) Pan Am G 61:369 (1934) (abst) G Soc Am, Pr 1934: 326 (1935)

34a West coast Eocene correlations. (abst) Pan Am G 42:77 (1934)

34b Santa Susana and lower Llajas fauna of Ventura County. (abst) Pan Am G 62:79 (1934) (abst) G Soc Am, Pr 1934:394 (1935)

35 Tectonics of the Mount Diablo and Coalinga areas, middle Coast Ranges of California. G Soc Am, B 46:1025-1078, 3 pls, 9 figs (1935) (abst) G Zentralbl, Abt A, Bd 57:372-373 (1936) (abst) Rv Geol et Sci conn 16:348-349 (1936)

36 (and **Vokes, H. E.**) Summary of marine Eocene sequence of western North America. G Soc Am, B 47:851-878, 2 pls, 3 figs (1936)

Clark, C. L.

33 (and **Kimberlin, C. L.**) Bottom hole pressure work at Kettleman. Petroleum World 1933 (April):19-20 (1933)

Clark, Douglas

33 Applications of geology to civil engineering. Cal Dp Nat Res, Div Mines, Cal M G, J, St Mineralogist's Rp 29: 161-173 (1933)

Clark, Frank R.

34 Origin and accumulation of oil. In Problems of petroleum geology. Am As Petroleum G, Pub, Sidney Powers Memorial Volume 309-335 (1934)

Clark, J. B.

33 (and **Young, W. H.**) Fuel briquets. U S B M, Min Yb 1934, Calendar Year 1933:645-652, California 648, 651 (1934)... 1935, Calendar Year 1934:711-718, California 715, 717 (1935)

36 (and **Plein, L. N.**) Fuel briquets. U S B M, Min Yb 1936:649-661, California 653, 657 (1936)

Clark, L. M.

31 Lower Miocene calcareous algae in California. Stanford Univ Micropaleontology B 3:15 (1931)

35 (and **Clark**, A.) The Vaqueros in the Temblor Range. (*abst*) Am As Petroleum G B 19:137 (1935)

Clark, Martha B.

32 Summary, mineral resources of the United States. U S B M, Min Res, Calendar Year 1929:A5-A120, California A95 (1932)... 1930:A5-A119, California A95 (1933)

34 (and **Shuey**, E. T.) Summary, mineral resources of the United States. U S B M, Min Res Calendar Year 1931: A5-A112, California A86 (1934)

34a (and **Shuey**, E. T.) Summary of mineral production. U S B M (Stat App) Min Yb 1932-1933:A1-A40, California A13 (1934)

35 Summary of mineral production. U S B M (Stat App) Min Yb 1934:A1-A41, California A14 (1935)... 1935:A1-A45, California A16 (1936)

Clark, Samuel G.

33 Milton formation of Sierra Nevada. (*abst*) Pan Am G 59:314-315 (1933) (*abst*) G Soc Am, Pr 1933 1:312-313 (1934)

Clark, W. A.

33 (**Wilhelm**, V. H., and **Davis**, E. L.) Characteristics of Edgewater encroachments in California fields. Oil Weekly 74, no 4:13-16 (1933) (*ann*) An Bib Ec G 6:294 (1934) M Metal 14:423-425 (1933) (*abst*) Rv Geol et Sci conn 15:345 (1935)

Clarke, F. W.

15 Analyses of rocks and minerals from the laboratory of the United States Geological Survey, 1880 to 1914. U S G S, B 591:376 pp (1915)

Clements, Frederick E.

36 (and **Chaney**, R. W.) Environment and life in the Great Plains. Carnegie Inst, Wash, Supplementary Pub 24:54 pp, California 16, 17, 18-19, 22, 25, 27, 42, 46, 48, figs, pls (1936)

Clements, Thomas

31 Structure of Tejon quadrangle. (*abst*) G Soc Am, B 42:315 (1931)

32 Extent of the Paleocene seas in the southerly part of the Tejon quadrangle, California. (*abst*) G Soc Am, B 43:226 (1932) (*abst*) Pan Am G 55:360-361 (1931)

33 Notes on the fall of columns during the Long Beach earthquake. Science 78: 100-101, 1 fig (1933)

34 (and **Oakeschott**, G. B.) Eocene (Martinez) of San Gabriel Mountains.

Clements, Thomas—Cont.

(*abst*) Pan Am G 61:307-308 (1934) (*abst*) G Soc Am, Pr 1934:310 (1935)

34a Some of the world's great gold mining districts. Pacific Mineralogist 1 (1): 3-4 (1934)

36 Bakersfield and petroleum. The Pacific Mineralogist 3, no 2:8-10 (1936)

Cloos, Ernst

31 (and **Balk**, R.) Primary structure of the Sierra Nevada intrusive in a cross section from Yosemite Valley to Mono Lake, California. (*abst*) N Jb 1931, Ref 2:311 (1931)

31a Mechanism of the intrusion of the granite masses between Mono Lake and the Mother Lode. (*abst*) G Soc Am, B 43:236 (1932) (*abst*) Pan Am G 55:273 (1931)

31b Der Sierra Nevada-Pluton. G Rundshau 22:372-384 (1931) (*abst*) Rv Geol et Sci conn 13:458-459 (1933) (*abst*) Am Geop Union, Tr 1935: Part I.:274 (1935)

32 Is the Sierra Nevada batholith a batholith? (*abst*) Wash Ac Sc, J 22:319-320 (1932)

32a (and **Johnston**, W. D. Jr) Structural history of the fracture systems of Grass Valley, California. Ec G 29:39-53 (1934) (*ann*) An Bib Ec G 7:41 (1935) (*abst*) Rv Geol et Sci conn 14:180 (1934) (*abst*) G Soc Am, B 44:88 (1933) (*abst*) G Zentralbl, Abt A 48:186-188 (1932) N Jb 1934, Ref 2: 523-524 (1934)... 156 (1935) (*abst*) Eng Index 1934:533 (1934) (*abst*) G Zentralbl, Abt A, Bd 53:348-349 (1936)

32b Structural survey of the granodiorite south of Mariposa, California. Am J Sc (5) 23:289-304, 3 figs (1932) (*abst*) Rv Geol et Sci conn 13:556-557 (1933) (*abst*) G Zentralbl, Abt A, 48:181-182 (1932)

33 Structure of the Sierra Nevada batholith. Int G Cong Guide Book 16: 40-45 (1933) (*abst*) N Jb 1934, Ref 2:523 (1934)

33a Sierra Nevada batholiths and the Mother-Lode. (*abst*) G Soc Am, B 44:79-80 (1933) J G 43:225-249 (1935) (*abst*) An Bib Ec G 1935:56 (1936)

33b Mother Lode and Sierra Nevada batholith. J G 43:225-249 (1935) (*abst*) Sc Progress 31:316 (1936)

Cloos, Hans

33 Uber Bau und Bewegung in Nordamerika. G Rundschau Bd 24:377-378 (1933)

Clute, Walker S.

36 Oil fields of the Bakersfield area. The Pacific Mineralogist 3, no 2:11-13 (1936)

Cockerell, Theodore Dru Alison

30 A fossil dragon fly from California (Odonata Calopterygidae) (*Protothore explicata* n. gen. and n. sp.) Entomological News 41:49-50, 1 pl (1930)

Cochrane, E W.

31 (and Hopkins, G R) Petroleum refineries in the United States January 1, 1931. U S B M, Inf Circ 6485:1-21, California 5, 6, 7-9 (1931)
32 (and Hopkins, G. R.) Petroleum refineries in the United States January 1, 1932. U S B M, Inf Circ 6641:1-21, California 3, 4, 6-9 (1932)
33 (and Hopkins, G. R.) Petroleum refineries, including cracking plants in the United States, January 1, 1933. U S B M, Inf Circ 6728:1-28 (1933)... January 1, 1934, Inf Circ 6807:1-30 (1934)

Cogen, William M.

36 Heavy mineral zones in the Modelo formation of the Santa Monica Mountains, California. Sedimentary Petrology J 6: 3-15, figs, maps (1936) (*abst*) Rv Geol et Sci conn 16:436 (1936)
36a Mechanics of the Lone Mountain landslides, San Francisco, California. Cal Dp Nat Res, Div Mines, M G, J, St Mineralogist's Rp 32:459-474 (1936)

Collom, R. E.

29 Oil accumulation and structure of Santa Maria district, California. (*abst*) Eng Index 1930:1300 (1930)
34 California oil reserves. Petroleum World An Rv 1934:26-31 (1934)

Colman, Murray N.

34 Treatment of tailings. Eng M J 135: 528-529 (1934)

Conant, L. C.

36 (Chelikowsky, J. R.; and Mayo, E. B.) Southern extension of the Mono Crater, California. Am J Sc 32 (5):81-97, figs, maps (1936)

Conkling, Harold

34 The depletion of underground water supplies: with discussion by W. P. Rowe. Am Geop Union, Tr (1934) pt 2:531-539, Nat Res Council (1934)

Coons, A. T.

32 Lime. U S B M, Min Res, Calendar Year 1929:267-298, California 267, 268, 271, 275, 277, 279, 281 (1932)... 1930:247-261, California 247, 248, 250, 252, 253, 255, 256, 258, 259, 261 (1932)... 1931:251-262, California 252, 254, 257, 258, 260 (1933)... Min Yb, Stat App 1932-1933:67-75, California 67, 69, 71, 74, 75 (1934)

Coons, A. T.—Cont.

32a Potash. U S B M, Min Res, Calendar Year 1929:139-145, California 140 (1932)... 1930:59-67, California 60 (1932)... 1931:23-32, California 24, 25 (1933)
32b Salt, bromine, and calcium chloride. U S B M, Min Res, Calendar Year 1929: 147-160, California 148, 149, 154, 155 (1932)... 1930:73-86, California 74, 75, 79, 80, 84, 85, 86 (1932)... 1931:45-60, California 46, 47, 50, 52, 53 (1933)... 1935:1029-1046, California 1030, 1031, 1032, 1033, 1035, 1036, 1045, 1046 (1935)
32c (and Bowles, O.) Slate. U S B M, Min Res, Calendar Year 1929:161-174, California 166 (1932)... 1930:277-290, California 283, 286, 290 (1932)... 1931:165-177 California 171, 173, 177 (1933) (Stat App) Min Yb 1932-1933:13-16, California 14 (1934)... 1936:833-840, California 837, 839 (1936)
32d Stone. U S B M, Min Res, Calendar Year 1929:229-266, California 233, 238, 239, 240, 242, 243, 245, 246, 250, 252, 259, 260, 261, 263, 264, 266 (1932)... 1930:333-373, California 337, 338, 343, 344, 345, 346, 348, 349, 351, 352, 356, 357, 358, 366, 367, 368, 369, 370, 371, 373 (1932)... 1931:297-329, California 301, 305, 306, 307, 309, 310, 312, 313, 314, 315, 322, 323, 324, 326, 327, 329 (1933)
32e (and Hopkins, G. R.) Crude petroleum and petroleum products. U S B M, Min Res, Calendar Year 1929:421-521, California 421, 423, 424, 430, 431, 432, 434, 435, 437, 439-444, 476, 477, 478, 481, 483, 484, 486, 488, 490, 500-503, 504, 507, 514, 516, 517, 518, 521 (1932)... 1930:775-876, California 777, 778, 779, 781, 782, 784, 785, 786, 788, 790, 792-796, 829, 830, 832, 823-834, 835, 836, 837, 838, 840, 841, 856-859, 860, 861, 862, 863, 868, 869, 871, 872, 873 (1932)... 1931:553-675, California 566, 567, 568, 570-571, 598, 599, 601, 607, 610, 612, 613, 614, 619, 622, 623, 624, 626, 627, 632, 633, 637, 638, 639, 640, 642, 643, 645, 647, 648, 654, 655, 656, 658, 659, 660, 661, 662, 665, 668, 671, 672, 673, 675 (1933)... Min Yb, Stat Yb, Stat App, 1932-1933: 299-372, California 303, 305, 306, 309, 311, 335, 336, 338, 339, 340, 341, 342, 345, 346, 348, 349, 350, 351, 353, 354, 359, 360, 361, 363, 364, 365, 369, 371, 372 (1934)... 1934: 199-281, California 200, 201, 203, 205, 206, 208, 209, 210, 213, 215, 230, 231, 232, 233, 234, 235, 236, 237, 238, 240, 242, 244, 245, 246, 247, 248, 250, 252, 253, 254, 255, 256, 257, 258, 259, 260, 262, 263, 268, 269, 270, 271, 272, 273, 274, 276, 277, 278, 279, 280 (1935)
32f (and Hatmaker, P.) Lime. U S B M, Min Yb 1932-1933: 629-637, California 631, 632 (1933)... 1934:861-871, California 863 (1934)

Coons, A. T.—Cont.

32g Natural sodium compounds and boron minerals. U S B M, Min Yb 1932-1933:795-798, California 797 (1933)... 1934: 1075-1078, California 1077 (1934)... 1935: 1187-1191, California 1188, 1189 (1935)... 1047-1050, California 1048, 1049, (1936)

32h Salt, bromine, calcium chloride, and iodine. U S B M, Min Yb 1932-1933:687-700, California 687, 688, 689, 691, 698, 699, 700 (1933)... 1934:929-945, California 930, 931, 932, 933, 934-935, 944 (1934)... 1936:915-930, California 916, 917, 918, 919, 920, 929 (1936)

32i Stone. U S B M, Min Yb Stat App 1932-1933:163-186, California 164, 168, 172, 174, 175, 176, 180, 181, 182, 183, 184, 185 (1934)

35 (and Bowles, O.) Limes. U S B M, Min Yb 935:967-976, California 969 (1935)

Cooper, Corwin L.

36 Mining and milling methods and costs at the Yellow Aster mine, Randsburg, Calif. U S B M, Inf Circ 6900:21 pp, 2 figs (1936)

Corey, W. H.

31 (and Loel, W.) Geologic history of the Vaqueros period in California. Petroleum World 28, no 8:55-77, 1 fig (1931)

32 (and Loel, W.) The Vaqueros formation, lower Miocene of California. I Paleontology. Cal Univ Dp G, B 22:31-410, 61 pls, 2 maps (1932) (abst) Rv Geol et Sci conn 13:523-524 (1933) Am J Sc 26:528-529 (1933) (abst) Eng Index 1933: 536 (1933) (abst) N Jb 1935, Ref 3:283-287 (1935)

35 Age of schist clastics, Venice district. Am As Petroleum G, B 19:1842 (1935) (ann) An Bib Ec G 8:353 (1936)

36 Age and correlation of schist-bearing clastics, Venice and Del Rey fields, California. Am As Petroleum G, B 20: 150-154 (1936) (abst) G Zentralbl, Abt A, Bd 57:51 (1936) (abst) Rv Geol et Sci conn 16:246 (1936)

Cornthwaite, M. A.

36 (and Johnson, B. L.) Barite and barium products. U S B M, Min Yb 1936:997-1006, California 998, 999 (1936)

36a (and Hughes, H. H.) Sand and gravel. U S B M, Min Yb 1936:841-848, California 844 (1936)

Corse, J. M.

31 (with Young, W. H.; and Tryon, F. G.) Fuel briquets. U S B M, Min Res, Calendar Year 1931:61-71, California 65, 67 (1933)

Corse, J. M.—Cont.

32 (and Young, W. H.) Fuel briquets. U S B M, Min Yb 1932-1933, Calendar Year 1932:451-458, California 456 (1933)

Crickmay, Colin H.

31 The anomalous stratigraphy of Deadman's Island, California. (abst) N Jb 1931, Ref 3:572 (1931) Biol Absts 5: 29103 (1931)

33 Mount Jura investigation. G Soc Am, B 44:895-926, 11 pls (1933) (abst) 44:80-81 (1933) (abst) Rv Geol et Sci conn 16:36 (1936)

33a Attempt to zone the North American Jurassic on the basis of its brachiopods. G Soc Am, B 44:871-894, 3 pls (1933)

Crook, T. H.

34 (and Kirby, J. M.) The Capay formation. (abst) Pan Am G 61:377 (1934) (abst) G Soc Am, Pr 1934:334-335 (1935)

Cross, C. M.

36 (and Hanna, G. D.; Taff, J. A.) Chico Cretaceous at the type locality. (abst) G Soc Am, Pr 1935:348 (1936)

Crown, W. J.

33 (and Pierce, G. G.; Howard, P. J.) Recent developments in the Long Beach oil field. California Oilfields 18, no 2:5-25, 4 maps (1933) (abst) Rv Geol et Sci conn 14:625 (1936) (ann) An Bib Ec G 7:100 (1935)

Cushman, Joseph Augustine

20 The American species of Orthophragmina and Lepidocyclina. U S G S pp 125D:41-42, pls, figs (1920)

25 Some Textulariidae from the Miocene of California. (abst) N Jb 1925, Ref 2, Band 2, Abt B:267 (1925)

31 (and Laiming, B.) Miocene foraminifera from Los Sauces Creek, Ventura County, California. J Paleontology 5:79-120, 5 figs, 6 pls (1931) (abst) Pale Zentralbl 3: 187 (1933) Biol Absts 9:457, entry 3895 (1935)

31a (and Parker, F. L.) Miocene foraminifera from the Temblor of the east side of the San Joaquin Valley, California. Cushman Lab Foram Res, Contr 7:1-16 (1931) (abst) Pale Zentralbl 2:354 (1933) Biol Absts 7:1216, entry 12025 (1933)

31b Foraminifera of Saratoga chalk. J Paleontology 5: 297-315, pls (1931)

31c (and Valentine, W. W.) Shallow water foraminifera from the Channel Islands of southern California. (abst) N Jb 1931, Ref 3:77 (1931)

Cushman, Joseph Augustine—Cont.

31d (and **Church,** C. C.) Some upper Cretaceous foraminifera from near Coalinga. (*abst*) N Jb 1931, Ref 3:76-77 (1931)

32 (and **Barbat,** W. F.) Notes on some arenaceous foraminifera from the Temblor formation of California. Cushman Lab Foram Res, Contr 8:29-40 (1932) (*abst*) Rv Geol et Sci conn 13:189 (1933) (*abst*) N Jb 1933, Ref 3:391 (1933) (*abst*) Pale Zentralbl 2:353-354 (1933)

32a Notes on the genus *Virgulina.* Cushman Lab Foram Res, Contr 8:7-23, pls (1932)

32b A bibliography of American foraminifera. Cushman Lab Foram Res, Sp Pub 3:1-40 (1932)

32c (and **Barksdale,** J. D.) Eocene foraminifera from Martinez, California. (*abst*) Rv Geol et Sci conn 12:632-633 (1932) Nature 128:1080 (1931) (*abst*) Pale Zentralbl 2:291 (1933)

34 (and **Campbell,** A. S.) A new Spiroplectoides from the Cretaceous of California. Cushman Lab Foram Res, Contr 10:70-71 (1934)

34a (and **Galliher,** E. W.) Additional new foraminifera from the Miocene of California. Cushman Lab Foram Res, Contr 10:24-26 (1934) (*abst*) N Jb 1934, Ref 3:624 (1934) (*abst*) Rv Geol et Sci conn 14:505 (1934)

34b (and **Kleinpell,** R. M.) New and unrecorded foraminifera from the California Miocene. Cushman Lab Foram Res, Contr 10:1-23 (1934) (*abst*) N Jb 1934, Ref 3:623-624 (1934) (*abst*) Rv Geol et Sci conn 14:504 (1934)

34c (and **Dusenbury,** A. N. Jr.) Eocene foraminifera of the Poway conglomerate of California. Cushman Lab Foram Res, Contr 10:51-65, pls (1934)

35 (and **Hobson,** J. A.) A foraminiferal faunule from the type San Lorenzo formation, Santa Cruz County, California. Cushman Lab Foram Res, Contr 11:53-64, pls (1935) (*abst*) Rv Geol et Sci conn 15: 606-607 (1935)

35a (and **Campbell,** A. S.) Cretaceous foraminifera from the Moreno shale of California. Cushman Lab Foram Res, Contr 11:65-73 pls, figs (1935) (*abst*) Rv Geol et Sci conn 15:606 (1936)

35b (and **Adams,** B. C.) New late Tertiary *Bolivinas* from California. Cushman Lab Foram Res, Contr 11:16-20, figs (1935)

35c (and **Siegfus,** S. S.) New species of foraminifera from the Kreyenhagen shale of Fresno County, California. Cushman Lab Foram Res, Contr 11:90-96 (1935)

35d (and **Hanna,** G. D.) Foraminifera from the Eocene near Coalinga, California. Biol Absts 9:457, entry 3894 (1935)

Dachnowski-Stokes, A. P.

33 Peat deposits in U. S. A. Handbuch der Moorkunde, Bd 7:1-140, 1-9 pls (1933) (*Review*) J Sed Petrology 4:100 (1934)

Daly, John W.

33 Paragenesis of mineral assemblages at Crestmore. (*abst*) Pan Am G 59:312-313 (1933) G Soc Am, Pr 1933 1:311 (1934) Am Mineralogist 20:638-659, 1 map, 1 table (1935) (*ann*) An Bib Ec G 8:265 (1936) (*abst*) G Zentralbl, Abt A, Bd 58:195 (1936) (*abst*) Rv Geol et Sci conn 16:260 (1936)

Dana, James Dwight

68 A system of Mineralogy. Fifth edition 827 pp (1868)

Darton, N. H.

34 Guidebook of the western United States. Part F. The Southern Pacific lines, New Orleans to Los Angeles. U S G S, B 845:304 pp, maps and illus (1934) (*abst*) N Jb 1934, Ref 2:350 (1934) (*Review*) Geog Rv 24:344-345 (1934) (*abst*) Rv Geol et Sci conn 14:204 (1934)

Davies, A. Morley

35 Tertiary faunas. Vol I. II. 685 pp, illus (1934, 1935)

35a A text book for oil field palaeontologists and students of geology. Vol I, II 658 pp, pls, illus (1935)

Davis, A. E.

30 (and **Hatmaker,** P.) Abrasive materials. U S B M, Min Res, Calendar Year 1930:151-169, California 152, 161, 163, 164 (1932)... 1931:111-130, California 112, 121, 124 (1933)

32 (and **Partridge,** E. P.) Magnesium and its compounds. U S B M, Min Yb 1932-1933:777-786, California 779 (1933)... 1934:1047-1056, California 1048, 1050 (1934)... 1935:1165-1176, California 1166, 1168 (1935)

34 (and **Bowles,** O.) Abrasive materials. U S B M, Min Yb 1934:889-906, California 891, 892 (1934)

35 Abrasive materials. U S B M, Min Yb 1935:995-1010, California 997, 999, 1000, 1001, 1002, 1005 (1935)

36 (and **Johnson,** B. L) Abrasive materials. U S B M, Min Yb 1936:877-894, California 878, 880, 883 (1936)

Davis, C. W.

32 (and **Davis,** H. W.) Platinum and allied metals. U S B M, Min Yb 1932-1933:337-345, California 337, 338 (1933)

Davis, E. L.

33 (**Wilhelm,** V. H.; and **Clark,** W. A.) Characteristics of edgewater encroachments in California fields. Oil Weekly

Davis, E. L.—Cont.

71, no 4:13-16 (1933) (*ann*) An Bib Ec G 6:294 (1934) M Metal 14:423-425 · (1933) (*abst*) Rv Geol et Sci conn 15:345 (1935)

Davis, Hubert W.

32 Platinum and allied metals. U S B M, Min Res, Calendar Year 1929:57-71, California 57 (1932)... 1930:99-112, California 99 (1933)... 1931:89-101, California 89, 97 (1934)... Min Yb 1934:507-516, California 508 (1934)... 1935:509-518, California 562 (1935)... 1936:509-518, California 510 (1936)

32a Iron ore, pig iron, and steel. U S B M, Min Res, Calendar Year 1929:5-44, California 7, 25 (1932)

33 (and **Davis,** C. W.) Platinum and allied metals. U S B M, Min Yb 1932-1933:337-345, California 337, 338 (1933)... 1935:561-580, California 562 (1935)

34 (and **Kiessling,** O. E.) Iron ore, pig iron, ferro-alloys, and steel. U S B M, Min Yb 1934:317-366, California 351 (1934)

35 Fluorspar and cryolite. U S B M, Min Yb 1935:1083-1105, California 1096-1097 (1935)

36 (and **Kiessling,** O. E., **Herring,** C. T.) Iron ore, pig iron, ferro-alloys and steel. U S B M, Min Yb 1936:365-394, California 375 (1936)

Davis, William Morris

31 Clear Lake, California. (*abst*) Science 74:572-573 (1931)

31a Shore lines of the Santa Monica Mountains, California. (*abst*) G Soc Am, B 43:227 (1932) (*abst*) Pan Am G 55:362-363 (1931)

31b (**Putnam,** W. C. and **Richards,** G. L.) Elevated shore lines of the Santa Monica Mountains, California. (*abst*) G Soc Am, B 42:309-310 (1931)

32 Glacial epochs of the Santa Monica Mountains, California. Nat Ac Sc, Pr 18:659-665, 8 figs (1932) G Soc Am, B 44: 1041-1133, 16 pls, 26 figs (1933) (*abst*) Pan Am G 59:306-307 (1933) (*abst*) G Soc Am, Pr 1933 1:304-305 (1934) (*abst*) Zs Glets-cherk 21:389 (*abst*) Rv Geol et Sci conn 16:31-32 (1936) (1934) (*Review*) Geog Rv 24:139 (1934) (*abst*) Nature 131:288 (1933)

32a Rock floors in arid and in humid regions. Zs Geom 7:250-253 (1932) (*abst*) Rv Geol et Sci conn 11:104-105 (1930)

33 California Journal of Mines and Geology. (*Review*) Am J Sc (5) 26:533 (1933)

33a Granitic domes of the Mojave Desert, California. San Diego Soc N H 7:211-258, 4 pls, 34 figs (1933) (*Review*) Geog Rv 24:138-139 (1934)

33b Mining in California. (*Review*) Am J Sc (5) 26:190 (1933)

Davis, William Morris—Cont.

33c San Francisco Bay. Int G Cong, Guide Book 16:18-21 (1933) (*abst*) N Jb 1936, Ref 3:66 (1936)

33d Submarine mock valleys. Am Geop Union, Tr 14:231-234 (1933) (*abst*) Pan Am G 59:307-308 (1933) (*abst*) G Soc Am, Pr 1933:306 (1934)

33e Work of sheet floods. (*abst*) G Soc Am, B 44:83 (1933)

33f The lakes of California. Cal Dp Nat Res, Div Mines, M G, J, St Mineralogist's Rp 29:175-236, 29 figs, 1 pl, map (1933) (*Review*) Geog Rv 24:138 (1934)

34 The Long Beach earthquake. Geog Rv 24 (1):1-11 (1934) (*abst*) Rv Geol et Sci conn 14:176 (1934)

35 (and **Maxson,** J. H.) Valleys of the Panamint Mountains, California. (*abst*) G Soc Am, Pr 1934:339 (1935)

Davison, Charles

34 The diurnal periodicity of earthquakes. J G 42:449-468 (1934)

Day, Arthur L.

25 (and **Allen,** E. T.) The volcanic activity and hot springs of Lassen Peak, California. (*abst*) Zs Vulkan 9:271-278 (1925-1926)

28 Report of the advisory committee in seismology. Carnegie Inst, Wash, Yb 27: 410-421 (1928) (*abst*) Canada, Dom Obs, Pub, Bib Seism 10:38 (1929)

32 Experiences of a seismologist with "seismic methods". Am Geop Union, 13th An Meeting, Tr:42-44 (1932) (*ann*) An Bib Ec G 5:191 (1932)

De Chardin, Teilhard P.

34 (and **Stirton,** R. A.) A correlation of some Miocene and Pliocene mammalian assemblages in North America and Asia with a discussion of the Mio-Pliocene boundary. Cal Univ, Dp G, B 23:277-290, 3 tables (1934) (*abst*) N Jb 1935, Ref 3:913 (1935)

Decius, L. Courtney

27 Natural gas development in California. Oil Gas J 26, no 4:75 (1927)

32 Contributions of petroleum geologists to general geology in California. (*abst*) Pan Am G 57:74-75 (1932)

Deflandre, Georges

34 Sur un foraminfere siliceux fossils des diatomites miocenes de Californie. Ac Sc, Paris, C R 198:1446-1448 (1934) (*abst*) Rv Geol et Sci conn 14:505 (1934)

Déribéré, Maurice

34 Le glucinium et ses derives. Revue des Produits chimiques 37:197-199, 228-230

Déribéré, Maurice—Cont.

(1934) (*abst*) Rv Geol et Sci conn 14:564-565 (1933-1934)

Dewell, H. D.

25 (and **Willis,** B.) Earthquake damage to buildings. Seism Soc Am, B 15:282-301, 15 pls (1925)

Dickerson, Roy E.

28 A criticism of the faunal relationships of the Meganos group by Bruce L. Clark. (*abst*) N Jb, Ref 1, Abt A, 1928: 362 (1928)

Diediker, Paul L.

30 (and **McDonald,** J. A.) A preliminary report on the foraminifera of San Francisco Bay, California. Stanford Univ, Micropaleontology B 2:33-38 (1930)

Diepenbrock, Alex

33 Mt. Poso oil field. Cal Dp Nat Res, Div Oil and Gas, California Oil Felds 19, no 2:5-35 (1934) (*ann*) An Bib Ec G 8: 154 (1935)

34 Round Mountain Field. Cap Dp Nat Res, Div Oil and Gas, California Oil Fields 19, no 4:5-19 (1934)

Dietrich, Waldemar Fenn

31 Clay prospecting and mining in California. (*abst*) N Jb 1931, Ref 2:600 (1931)

Doane, George H.

36 California dry hole record. Petroleum World, An Rv 1936:123-178 (1936)

Dodd, H. V.

30 Controlling a blow-out in the Kettleman field. Petroleum World (London) 27:223-228 (1930)

31 Recent developments in the Kettleman Hills field. Cal Dp Nat Res, Div Oil and Gas, California Oil Fields 17, no 1:5-44 (1931) (*ann*) An Bib Ec G 5:341 (1932) (*abst*) Eng Index 1932:899 (1932) (*abst*) Rv Geol et Sci conn 13:235 (1933) (*abst*) G Zentralbl 51:255 (1934)

31a Operations in district No. 5, 1930. Cal Dp Nat Res, Div Oil and Gas, California Oil Fields 16, no 3:66-73 (1931)... 1931, 17, no 3:50-55 (1932)... 1932, 18, no 3:49-56 (1933)... 1933, no 3:50-56 (1934)

33 (and **Kaplow,** E. J.) Kettleman North Dome and Kettleman Middle Dome fields—progress in development. Cal Dp Nat Res, Div Oil and Gas, California Oil Fields 18, no 4:5-20 (1933 (*ann*) An Bib Ec G 7:324 (1934)

Doerner, H. A.

30 Roasting of chromite ores to produce chromates. U S B M, Rp Invest 2999: 1-30 (1930)

Doerner, H. A.—Cont.

30a Concentration of chromite. U S B M, Rp Invest 3049:1-8, California 2, 3, 4-5 (1930)

Dolbear, Samuel H.

34 Magnesite. Min Ind 42:380-386, California 380-382 (1934)... 43:490-395, California 390, 391 (1935)

Dolman, S. G.

31 Elwood oil field, California. Cal Dp Nat Res, Div Oil and Gas, California Oil Fields 16, no 3:5-12, 4 pls (1931) (*abst*) Rv Geol et Sci conn 13:234 (1933) (*abst*) Eng Index 1932:899 (1932) (*ann*) An Bib Ec G 8:154 (1935) California Oil Fields 16:484-487 (1931)

31a Operations in district No. 3, 1930. Cal Dp Nat Res, Div Oil and Gas, California Oil Fields 16, no 3:43-51 (1931)... 1931, 17, no 3:33-39 (1932)... 1932, 18, no 3:33-40 (1933)

32 Lompoc oil field, Santa Barbara County, California. Cal Dp Nat Res, Div Oil and Gas, California Oil Fields 17, no 4:13-20 (1932)

Donnay, J. D. H.

31 Genesis of the Engels copper deposit; a field study and microscopic investigation of a late magmatic deposit. Congrès Intern Mines, Mét et Géol appl. 4:99-111 (1930) (*ann*) An Bib Ec G 4:233 (1931) (*abst*) Rv Geol et Sci conn 14: 69-70 (1934)

Donnelly, Maurice

34 Economic geology of Julian region. (*abst*) Pan Am G 61:316 (1934) (*abst*) G Soc Am, Pr 1934:321 (1935)

34a Geology and mineral deposits of the Julian district, San Diego County, California. Cal Dp Nat Res, Div Mines, M G J, St Mineralogist's Rp 30:331-370, map (1934) (*ann*) An Bib Ec G 8:55 (1935) (*abst*) G Zentralbl, Abt A, Bd 57:156 (1936)

35 Orthoclase from San Diego County pegmatites. (*abst*) Pan Am G 63:320 (1935) (*abst*) G Soc Am, Pr 1935:341 (1936)

36 Notes on the lithium pegmatites of Pala, California. The Pacific Mineralogist 3, no 1:8-12 (1936)

Dorf, Erling

33 (and **Chaney,** R. W.) Ecology of the Tertiary forests of western North America. (*abst*) G Soc Am, Pr 1933 1:357 (1934)

33a Pliocene floras of California. Carnegie Inst, Wash, Pub 412:1-112, 13 pls, 1 fig (1933) Biol Absts 7:2246, entry 22152 (1933)

Dorn, C. L.

32 Report on a deep boring in Salinas Valley, California. Stanford Univ Micropaleontology, B 3:28-29 (1932)

Dreyer, F. E.

35 Geology of Mount Pinos quadrangle. (abst) Pan Am G 64:74-75 (1935) (abst) G Soc Am, Pr 1935:351 (1936)

Driggs, F. H.

31 (and **Marden, J. W.**) Titanium and zirconium. Min Ind 39:602-611, California 603-604 (1931)

Dudley, Paul H.

31 Geology of a portion of the Perris block, southern California. (abst) G Soc Am, B 43:223 (1932) (abst) Pan Am G 55:358 (1931) Cal Dp Nat Res, Div Mines, M G, J, St Mineralogist's Rp 31:487-507 (1935)

36 Physiographic history of a portion of the Perris block, southern California. J G 44:358-378 (1936) (abst) G Zentralbl Abt A, Bd 57:510 (1936) (abst) Rv Geol et Sci conn 16:368 (1936)

Dunham, K. C.

33 A note on the texture of the Crestmore contact rocks. Am Miner, J 18:475-477 (1933) Min Absts 5:443 (1934)

33a (and **Larsen, E. S.**) Tilleyite, a new mineral from the contact zone at Crestmore, California. Am Miner, J 18:469-473 (1933) Min Absts 5:387 (1934) (abst) N Jb 1934, Ref 1:142 (1934)

Dunlop, J. P.

32 Gold and silver. U S B M, Min Res, Calendar Year 1929:877-920, California 877, 885, 886, 891, 892, 896, 897, 900, 901, 904, 905, 907, 908, 909, 912 (1932)... 1930:817-857, California 827, 828, 830, 831, 833, 834, 835, 836, 838, 839, 840, 841, 843, 844, 845, 846, 847, 848, 850, (1933)... 1931: 679-710, California 680, 686-687, 688, 689, 691, 692, 693, 694, 695, 698, 699, 701, 702, 703, 704, 705-706, (1934) U S B M, Min Yb Stat App 1932-1933:455-479, California 455, 462, 463, 464, 465, 467, 468, 469, 470, 471, 472, 473, 474, 475, 476, 477, 478, (1934)... 1934:335, 337-351 (1935)... 1935:327-352, California 327, 335, 337, 338, 340, 341, 343, 344, 345, 346, 348 (1936)

36 (and **Henderson, Chas. W.**) Gold and silver. U S B M, Min Yb 1936:91-105, California 92, 93, 95, 101, 103 (1936)

Dunn, J. A.

33 Andalusite in California and kyanite in North Carolina. Ec G 28:692-695 (1933) (ann) An Bib Ec G 6:244 (1934)

Dusenbury, Arthur N. Jr.

32 A faunule from the Poway conglomerate. Upper middle Eocene of the San Diego County, California. Stanford Univ Micropaleontology B 3:84-95 (1932)

34 (and **Cushman, J. A.**) Eocene foraminifera of the Poway conglomerate of California. Cushman Lab Foram Res, Contr 10:51-66, pls (1934)

Dyk, Karl

32 (and **Byerly, P.**) Richmond quarry blast of September 12, 1931 and the surface layering of the earth in the region of Berkeley. Seism Soc Am, B 22:50-55 (1932)

Dyk, Robert

28 (and **Byerly, P.**) Registration of earthquakes at the Berkeley Station and at the Lick Observatory Station from October 1, 1927 to March 31, 1928. Cal Univ, Seism Sta, B 2:273-300 (1928)

30 (and **Byerly, P.**) Registration of earthquakes at the Berkeley Station and the Lick Observatory Station from October 1, 1929 to March 31, 1929. Cal. Univ. Seism Sta, B 2:399-439 (1930)

30a (and **Byerly, P.**) Registration of earthquakes at the Berkeley Station and at the Lick Observatory Station from April 1, 1929 to September 30, 1929. Cal Univ, Seism Sta, B 2:361-397 (1930)

Dykes, Leland H.

32 Occurrence of monazite in a granodiorite pegmatite (Riverside County, California) (abst) Pan Am G 58:74 (1932) (abst) G Soc Am, B 44:161 (1933) (ann) An Bib Ec G 6:72 (1934)

Eardley-Wilmot, V. L.

31 Abrasives. Min Ind 39:1-12, California, Pumice 7 (1931)... 40:1-14, California 1, 4, 7 (1932)... 41:1-12, California 1, 6-7 (1933)... 42:1-12, California 7 (1934)... 43: 1-12, California 3, 6 (1935)

31a Diatomite. Min Ind 39:213-220, California 215-216 (1931)... 40:173-179 California 175 (1932)... 41:164-168, California 165-166 (1933)... 42:177-183, California 178-179 (1934)... 43:175-179, California 176 (1935)

Eaton, J. Edmund

31 Standards in correlation. Am As Petroleum G, B 15:367-384, 4 figs (1931)

32 Decline of the Great Basin, southwestern United States. Am As Petroleum G, B 16:1-49, 10 figs (1932) (abst) G Zentralbl, Abt A 47:188 (1932) N Jb 1933, Ref 3:575-577 (1933)

Eaton, J. Edmund—Cont.

33 Clastic facies and faunas of Monterey formation, California. Am As Petroleum G, B 17:1009-1015 (1933) (*ann*) An ,Bib Ec G 6:294-295 (1934) (*abst*) G Zentralbl 51:56 (1934) (*abst*) Eng Index 1934:516 (1934)

33a Geology of the southern California earthquake. Petroleum World 30, no 4:13-14 (1933)

33b Long Beach, California, earthquake of March 10, 1933. Am As Petroleum G, B 17:732-738 (1933) (*abst*) Canada, Dom Obs, Pub, Bib Seism 10:343 (1933) (*abst*) Eng Index 1933:774 (1933)

35 Miocene of Caliente Range, California. Am As Petroleum G, B 19:1844 (1935)

35a California is earth's youngest child. Petroleum World 1935 (November):89-93, 100 (1935)

35b Outlook for new fields in California. Petroleum World An Rv 1935:29-34 (1935)

Eckel, E. C.

34 Limestone deposits of the San Francisco region. Cal Dp Nat Res, Div Mines and Mining, M G, J 29:348-361, (1933) (*ann*) An Bib Ec G 7:58 (1935) (*abst*) Eng Index 1934:639 (1934)

Eckis, Rollin

31 Alluvial fans of Cucamonga district, southern California. (*abst*) N Jb 1931, Ref 3:932 (1931)

34 South Coastal Basin investigation. Geology and ground water storage capacity of valley fill. Cal Dp Pub Wks, Div Water Resources, B 45:1-273 (1934) (*ann*) An Bib Ec G 8:375 (1936)

35 Late Quaternary geology of Los Angeles basin. (*abst*) Pan Am G 64:74 (1935) (*abst*) G Soc Am, Pr 1935: 350 (1936)

35a (and Gross, P. L. K.) Porosity and sorting of California fanglomerates. (*abst*) Pan Am G 64:77 (1935) (*abst*) G Soc Am, Pr 1935:353 (1936)

Edwards, Everett C.

34 Pliocene conglomerates of Los Angeles Basin and their paleogeographic significance. Am As Petroleum G, B 18:786-812, 7 figs (1934) (*abst*) Rv Geol et Sci conn 14:453 (1934) Inst Pet Tech, J 20: 435A (1934) (*abst*) Eng Index 1934:802 (1934) (*abst*) G Zentralbl, Abt A, Bd 53: 314 (1934)

Edwards, M. G.

33 Some Eocene localities in Salinas Valley district, California. Am As Petroleum G, B 17:81 (1933)

Edwards, S. C.

34 Sand concretions from California. Rocks & Min 9:82-83 (1934)

Effinger, W. L.

35 Gaviota formation of Santa Barbara County. (*abst*) Pan Am G 64:75-76 (1935) (*abst*) G Soc Am, Pr 1935:351-352 (1936)

Eggleston, H. L.

34 The outlook for natural gasoline. Petroleum World An Rv 1934:144-147, 166 (1934)

Ellsworth, E. W.

33 Physiographic history of Afton basin of Mojave Desert. (*abst*) Pan Am G 59:308-309 (1933) (*abst*) G Soc Am, Pr 1933 1:306-307 (1934)

33a Tracing buried river channels by geomagnetic methods. Cal Dp Nat Res, Div Mines, M G, J, St Mineralogist's Rp 29:244-250 (1933) (*ann*) An Bib Ec G 6:335 (1934)

36 (and Blackwelder, E.) Pleistocene lakes of the Afton basin. Am J Sc 31 (5):453-463, figs (1936)

Emery, Alden H.

33 (and Stoddard, B. H.) Talc and ground soapstone. U S B M, Min Yb 1932-1933:715-722, California 717, 718, 719 (1933)... 1934:975-984, California 976, 977-978 (1934)... 1935:1069-1081, California 1070, 1071, 1072 (1935)... 1936:953-962, California 953, 954, 955 (1936)

Emmons, William Harvey

31 Geology of petroleum. (2nd edition) 736 pp, figs, California 520-565. Petrolia 524. Petaluma 524. Santa Clara and Sargent 524. Coalinga 525-528. Kettleman Hills 528-529. Devil's Den 530. Lost Hills 530. Buttonwillow 530. North Belridge 531. Belridge 531. McKittrick, Sunset, and Midway 531-537. Buena Vista Hills 537. Elk Hills 538-539. Kern River 539-541. Arroyo Grande 541. Santa Maria 541-544. Goleta 544. Elwood 544. Summerland 544. Santa Clara Valley 545-551. Los Angeles Basin 551-565 (1931)

Engel, Rene

31 Geology of the southwest quarter of the Elsinore quadrangle. (*abst*) G Soc Am, B 43:225 (1932) (*abst*) Pan Am G 55:360 (1931)

34 Geochemical relations between waters in Elsinore region. (*abst*) Pan Am G 61:317 (1934)

Engeln, O. D. von

32 Ubehebe craters and explosion breccias in Death Valley, California. J G 40:726-734 (1932) (*abst*) Eng Index 1933:536 (1933) (*abst*) Zs Vulkan 15:209 (1933) Rv Sc Progress 29:121 (1934) (*abst*) N Jb 1934, Ref 2:563-564 (1934) (*abst*) Rv Geol et Sci conn 14:21 (1933-1934)

33 Palisade glacier of the high Sierras of California. G Soc Am, B 44:575-599 (1933) (*abst*) Zs Gletscherk 21:388 (1934) (*abst*) Rv Geol et Sci conn 15:549 (1935)

Engineering and Mining Journal

31 California moves to stabilize mineral production. Eng M J, June 8 (1931)

34 Eighty-six years of gold production in California. Eng M J 135:483-485 (1934) (*ann*) An Bib Ec G 7:243 (1934)

34a California gold continues to enrich the nation. Eng M J 135:481 (1934)

34b Equipment of gold mines and mills in California. Eng M J 135:522-527 (1934)

Engineering News-Record

33 Destructive earthquake centers on Long Beach, California. Eng News-Rec 110:353-355 (1933) (*abst*) Eng Index 1933: 327 (1933)

35 Underground water storage in California's south coastal basin. Eng News-Rec 115:733-738 (1935) (*abst*) An Bib Ec G 8:376 (1936)

Erich, E. E.

31 Mining opportunities in known districts. Geologic map of Weaverville quadrangle. (*abst*) N Jb 1931, Ref 2:435 (1931)

Erwin, Homer D.

34 Geology and mineral resources of northeastern Madera County, California. Cal Dp Nat Res, Div Mines, M G, J, St Mineralogist's Rp 30:7-78 (1934)

34a Notes on sampling as applied to gold quartz deposits. Cal Dp Nat Res, Div Mines, M G, J, St Mineralogist's Rp 30:91-99 (1934)

36 (and **Averill**, Chas. Volney) Mineral resources of Lassen County. Cal Dp Nat Res, Div Mines, M G, J, St Mineralogist's Rp 32:405-444 (1936)

Estorff, Fritz E., von

31 Kreyenhagen shale at type locality, Fresno County, California. (*abst*) Rv Geol et Sci conn 11:372-373 (1930) (*abst*) G Zentralbl 44:58 (1931) Biol Absts 6:567, entry 5255 (1932)... 6:1186, entry 11262 (1932) (*abst*) Pale Zentralbl 2:222 (1933)

33 (and **Barbat**, W. F.) Lower Miocene foraminifera from southern San Joa-

Estorff, Fritz E., von—Cont.

quin Valley, California. J Paleontology 7: 164-174, 1 pl, 1 fig (1933) (*abst*) N Jb 1934, Ref 3:422 (1934) Pale Zentralbl 5:271-272 (1934) Biol Absts 9:916, entry 8195 (1935)

Etcheverry, Bernard Alfred

33 (and others) Report on Iron Canyon, Table Mountain, and Kennett dam sites, on Sacramento River. Cal Dp Pub Works, Div Water Res, B no 26 (1931):455-462 (1933)

Evans, R. D.

35 (and **Williams**, H.) The radium content of lavas from Lassen Volcanic National Park, California. Am J Sc (5) 29: 441-452 (1935)

Fairchild, J. G.

20 (and **Miser**, H. D.) Hausmannite in the Batesville district, Arkansas. Wash Ac Sc J 10:1-8 (1920)

32 (and **Schaller**, W. T.) Bavenite, a beryllium mineral, pseudomorphous after beryl, from California. Am Mineralogist 17:409-422 (1932) (*ann*) An Bib Ec G 5:255 (1932) (*abst*) N Jb 1933, Ref 1:235-237 (1933) Min Absts 5:230-231 (1933) (*abst*) Rv Geol et Sci conn 14:97 (1934) Chem Absts 26:5515 (1932)

Farnsworth, H. R.

32 (**Woodring**, W. P.; and **Roundy**, P. V.) Geology and oil resources of the Elk Hills, California (including Naval Petroleum Reserve No 1) U S G S, B 835:82 pp, 8 figs, 22 pls (1932) (*abst*) Rv Geol et Sci conn 13:367, 513 (1933) (*abst*) Eng Index 1933:818 (1933)

Farquhar, Francis P. (Editor)

31 Up and down in California. The journal of William H. Brewer. Yale Univ Press: 601 pp, 32 pls (1930) (*abst*) Am J Sc 21:466-467 (1931) (*Review*) Nature 127: 968 (1931)

35 The story of Mount Whitney, part 2. Sierra Club, B 20:81-94 (1935)... Part 3, 1936:64-75 (1936)

Faustino, Leopoldo

31 Two new madreporarian corals from California. Philippine J Sc 44, no 3:285-289, 1 pl (1931) Biol Absts 7:721, entry 7024 (1933) (*abst*) G Zentralbl 45:334 (1931) Pale Zentralbl 2:103 (1932)

Fenneman, Nevin M.

32 Physiography of western United States. (*abst*) Zs Geom Band VII, Heft 4/5:265-266 (1932)

Fenner, C. N.

31 The Engels copper deposits, California. (*abst*) G Zentralbl 43:408 (1930-1931) Chem Absts 25:1464 (1931)

Ferguson, Henry G.

30 Vein quartz of the Alleghany district, California. (*abst*) Wash Ac Sc, J 20:151-152 (1930)

32 (and **Gannett,** R. W.) Gold quartz veins of the Alleghany district, Californit. U S G S, P P 172:139 pp (1932) Am I M Eng, Tech Pub 211 (Class 1, mining geology 24):40 pp, 17 figs (1929) (*abst*) M Metal 10:252 (May 1929) (*abst*) Rv Geol et Sci conn 12:88-90 (1932) (*ann*) An Bib Ec G 5:54 (1932) (*abst*) G Zentralbl, Abt A, 48:362-363 (1932) (*abst*) N Jb 1933, Ref 2:186-187 (1933) (*abst*) Eng Index 1932:632 (1932) Chem Absts 26:4776 (1932) (*abst*) Rv Geol et Sci conn 13:587 (1933) (Review by **Knopf,** A.) Ec G 28:399-401 (1933)

Finch, Ruy H.

26 (and **Jaggar,** T. A.) Tilting and level changes at Pacific volcanoes. Third Pan-Pacific Sc Cong, Tokyo, Pr 1:672-686 (1926) (*abst*) Canada Dom Obs, Pub, Bib Seism 10:42 (1929)

30 Activity of California volcano in 1786. The Volcano Letter, no 308: November 21 (1930)

31 Some work of the Lassen volcano observatory. The Volcano Letter, no 336: 1-3 June 4 (1931)

31a The youngest lava flow on the mainland of the United States. The Volcano Letter, no 334: May 21 (1931)

31b (and **Anderson,** C. A.) Quartz-basalt eruptions of Cinder Cone. (*abst*) N Jb 1931, Ref 2:277 (1931) G Zentralbl 46:232-233 (1932) Zs Vulkan 14:94-95 (1931)

31c Rainfalls accompanying explosive eruptions of volcanoes. F P S C, Pr 2 B: 511-516 (1930) (*abst*) Zs Vulkan 14:94 (1931)

32 Work of the Lassen volcano observatory in 1931. The Volcano Letter, no 373: 1-2 February 18 (1932)

33 Block lava. J G 41:769-770 (1933) (*abst*) N Jb 1934, Ref 2:552 (1934)

33a Burnt lava flow in northern California. Zs Vulkan 15:180-183, pls 8-9, 3 figs (1933) (*abst*) N Jb 1935, Ref 2:492 (1935)

33b Slump scarps. J G 41:647-649 (1933) (*abst*) N Jb 1934, Ref 2:552 (1934)

33c Lassen report, no 32. The Volcano Letter, no 395:1 January (1933) (*abst*) Zs Vulkan 15:196 (1933)

Findlay, W. A.

33 (and **Bode,** F. D.) Structure of a part of the San Joaquin Hills, California.

Findlay, W. A.—Cont.

(*abst*) Pan Am G 59:318-319 (1933) (*abst*) G Soc Am, Pr 1933 1:316 (1934)

Fink, Colin G.

31 Tungsten. Min Ind 39:612-623, California 614 (1931)... 40:555-566, California 557-558 (1932)... 41:525-535, California 527 (1933)... 42:571-581, California 571, 572 (1934)... 43:570-583, California 572 (1935)

Fisher, E. H.

31 Natural gas aids in sand production. Am Gas J 134:39-40, 7 figs (1931) (*abst*) Eng Index 1931:1255 (1931)

Fisher, Edna M.

30 Early fauna of Santa Cruz Island, California. J Mammalogy 11, no 1:75-76 (1930) Biol Absts 7:760, entry 7492 (1933)

30a Early fauna of the Santa Barbara region, California. J Mammalogy 11, no 2:223-224 (1930) Biol Absts 7:243, entry 2442 (1933)

Fisher, H. F.

31 Microscopic study of California oil field emulsions and some notes on the effects of superimposed electrical fields. Am I M Eng, Petr Div, Tr, Petroleum Dev and Tech 1931:359-375 (1931) (*ann*) An Bib Ec G 5:134 (1932)

Fitch, A. A.

31 Barite and witherite from near El Portal, Mariposa County, California. Am Mineralogist 16:461-468, 3 figs (1931) (*ann*) An Bib Ec G 4:272 (1931) (*abst*) N Jb 1931, Ref 1:554-555 (1931) Min Absts 5:66 (1932)

31a The geology of Ben Lomond Mountain, California. Cal Univ Dp G, B 21:1-13, 2 figs (1931) (*abst*) G Zentralbl, Abt A 48:507-508 (1933) (*abst*) Eng Index 1932: 612 (1932)

32 A contact section on the Mokelumne River, California. Cal Univ Dp G, B 22: 1-12, 2 figs, 1 pl (1932) (*abst*) Eng Index 1932:612 (1932)

32a The Sierra Nevada as a comagmatic region. Am J Sc (5) 24: 481-495, 5 figs (1932) (*Review*) Sc Progress 29:120 (1934)

Foote, H. W.

10 (and **Langley,** R. W.) On an indirect method for determining columbium and tantalum. Am J Sc (4) 30:393-400 (1910)

Forbes, Hyde

29 Geology of damsites on American River. Cal Dp Pub Works, Div Water Resources, B 24:175-190 (1929)

Forbes, Hyde—Cont.

31 Geology and underground water storage capacity of Sacramento valley. Cal Dp Pub Works, Div Water Resources, B 26:512-532 (1933)

31a Geologic reports on damsites in Sacramento River Basin. Cal Dp Pub Works, Div Water Resources, B 26:479-514 (1933)

31b Geology and underground water storage capacity of San Joaquin Valley. Cal Dp Pub Works, Div Water Resources, B 29:530-550 (1931)

31c Geological reports on damsites in San Joaquin River Basin. Cal Dp Pub Works, Div Water Resources, B 29:552-606 (1931)

Foshag, William Frederick

25 Famous mineral localities: Furnace Creek, Death Valley, California. (abst) N Jb 1925, Band 2, Abt A 1925:299 (1925)

31 A new sulfate of iron and potash from California. (abst) Am Mineralogist 16:115 (1931)

31a Krausite, a new sulphate from California. Am Mineralogist 16:352-360 (1931) (ann) An Bib Ec G 5:34 (1932) (abst) N Jb 1932:45-46, Band 1 (1932) Min Absts 5:51 (1932) Chem Absts 25:5644-5645 (1931)

31b Schairerite, a new mineral from Searles Lake, California. Am Mineralogist 16:133-139, 5 figs (1931) (abst) N Jb 1931, Ref 1:331-333 (1931) Min Absts 4:498 (1931) Chem Absts 25:5116 (1931)

31c Probertite from Ryan, Inyo County, California. Am Mineralogist 16:338-341 (1931) (ann) An Bib Ec G 5:35 (1932) Min Absts 5:52 (1932) Chem Absts 25:5644 (1931)

32 (and Palache, C.) The chemical nature of joaquinite. Am Mineralogist 17:308-312, 1 fig (1932) Min Absts 5:191 (1932)

35 Burkeite, a new mineral species from Searles Lake, California. Am Mineralogist 20:50-56 (1935) (ann) An Bib Ec G 8:26 (1935) (abst) N Jb 1935, Ref 1:144 (1935) Chem Absts 29:6176 (1935)

36 (and Woodford, A. O.) Bentonitic magnesian clay-mineral from California. Am Mineralogist 21:238-244 (1936) (abst) Rv Geol et Sci conn 16:435-436 (1936)

Fourmarier, P.

33 Observations sur la Geologie et quelques Cites mineraux de L'Amerique du Nord. Rv Univ Mines (8) 9:61-67 Sierra Nevada 67 (1933)

Fowler, H. C.

32 Technical developments in petroleum and natural gas production. U S B M, Min Yb 1932-1933:497-509, California 507 (1933)

Fowler, H. C.—Cont.

35 (and Kennedy, H. S.) Natural gas. U S B M, Min Yb 1935:795-819, California 798-801 (1935)

35a (and Cattell, R. A.) Influence of technology upon composite interest in oil. U S B M, Min Yb 1935:771-794, California 773, 778, 789-791, 793-794 (1935)

Franke, Herbert A.

30 Santa Clara County. (abst) Eng Index 1930:1114-1115 (1930)

30a San Francisco field division: Kings County. Cal Dp Nat Res, Div Mines, Mining in California, St Mineralogist's Rp 26:413-471 (1931)

31 Tulare County. Cal Dp Nat Res, Div Mines, Mining in California, St Mineralogist's Rp 26:423-471 (1931)

32 Selected bibliography on placer mining. Cal Dp Nat Res, Div Mines, Mining in California, St Mineralogist's Rp 28:219-224 (1932)

35 Mines and mineral resources of San Luis Obispo County. Cal Dp Nat Res, Div Mines, M G, J, St Mineralogist's Rp 31:402-461, tables, figs, maps (1935)

35a Mineral resources of portions of Monterey and Kings County. Cal Dp Nat Res, Div Mines, M G, J 31:462-465 (1935)

36 (and Logan, C. A.) Mines and mineral resources of Calaveras County. Cal Dp Nat Res, Div Mines, M G, J, St Mineralogist's Rp 32:226-364 (1936)

Franz, Shepherd Ivory

34 (and Norris, A.) Human reactions to the Long Beach earthquake. Seism Soc Am, B 24:109-114 (1934)

Fraser, Donald McCoy

31 Geology of San Jacinto quadrangle south of Gorgonio Pass, California. Cal Dp Nat Res, Div Mines, Mining in California, St Mineralogist's Rp 27:494-540, 22 figs (1931) (abst) G Soc Am, B 42:235 (1931) (abst) Pan Am G 55:318-319 (1931) (abst) Eng Index 1932:612 (1932)

Freeman, J. R.

32 Earthquake damage and earthquake insurance. McGraw Hill Book Co., N. Y.: 904 pp (1932) (Review) Nature 131:381-382 (1933)

Frost, F. H.

34 The Pleistocene flora of Rancho La Brea. (abst) N Jb 1934, Ref 3:359 (1934)

Furlong, Eustace L.

31 Capromeryx minor Taylor from the McKittrick Pleistocene, California. (abst) G Zentralbl 44:68 (1931) Pale Zentralbl 3:440 (1933)

Furlong, Eustace L.—Cont.

31a Distribution and description of skull remains of the Pliocene antelope Sphenophalos from the Northern Great Basin Province. Carnegie Inst, Wash, Pub 418: 29-36, 1 fig, 5 pls (1931)

Furness, J. W.

35 (and **Meyer,** H. M.) Mercury. U S B M, Min Yb 1935:437-463, California 442, 449-451 (1935)

35a (and **Pehrson,** E. W.; **Meyer,** H. M.) Copper. U S B M, Min Yb 1935:45-73, California 51, 53, 54, 55, 56 (1935)... 1936: 107-135, California 112, 114, 116-118 (1936)

Gale, Hoyt Rodney

31 Correlation between later Cenozoic deposits of California and of Europe. Am As Petroleum G, B 15:555-556 (1931) (*abst*) G Zentralbl 45:253 (1931)

31a Summary of stratigraphy. *In* Grant and Gale, catalogue of the marine Pliocene and Pleistocene mollusca of California. San Diego Soc N H, Mem 1:19-78 (1931)

31b (and **Gale,** H. S.) Miocene vulcanism. (*abst*) Pan Am G 55:371 (1931) G Soc Am, B 43:234-235 (1932)

31c (and **Grant,** U. S.) Catalogue of the marine Pliocene and Pleistocene mollusca of California. San Diego Soc N H, Mem 1:1036 pp, 15 figs, 32 pls (1931) (*abst*) G Zentralbl Abt A 47:393-395 (1932) N Jb 1933, Ref 3:168-174 (1933) Nature 129:135 (1932) Pale Zentralbl 2:186-190 (1933)

33 Summary of west coast subgenus *Trophosycon* (*abst*) Pale Zentralbl 2:373 (1933)

Gale, Hoyt S.

31 (and **Gale,** H. R.) Miocene vulcanism. (*abst*) Pan Am G 55:371 (1931) (*abst*) G Soc Am, B 43:234-235 (1932)

31a (and **Scofield,** C. S.) McKenzie Taylor's genesis of petroleum and coal as applied to Fruitvale field, California. Am As Petroleum G, B 15:709-712 (1931)

32 (and others) Geology of southern California. Int G Cong, Guide Book 15: 1-10 (1932) (*abst*) N Jb 1936, Ref 3:55-56 (1936) (*abst*) G Zentralbl, Abt A, Bd 53: 240 (1934)

32a (**Piper,** A. M.; and **Thomas,** H. E.) Geology of the Mokelumne River basin, California. U S G S typescript Rp 377 pp (1932) (*abst*) Nat Res Council, B 98: 192-193 (1935)

33 Hopeful view of Huntington Beach tide lands. Cal Oil World 26, no 21:11-12 (1933)

34 Geology of Huntington Beach oil field, California. Am As Petroleum G,

Gale, Hoyt S.—Cont.

B 18:327-342 (1934) (*ann*) An Bib Ec G 7:100 (1935) (*abst*) Rv Geol et Sci conn 14:492-493 (1934) Inst Pet Tech, J 20:. 268A-269A (1934) (*abst*) Eng Index 1934: 802 (1934)

34a Real field lies in the sea at Huntington. Petroleum World 1934 (June):16-18 (1934)

Galiher, C.

32 (and **Adams,** W. W.) Fuller's earth. U S B M, Min Yb 1932-1933:709-714, California 709, 711, 712, 713 (1933)

32a (and **Metcalf,** R. W.) Gypsum. U S B M, Min Yb 1932-1933:17-23, California 18, 20 (1933)

36 (and **Rogers,** H. O.) Feldspar. U S B M, Min Yb 1936:735-743, California 740, 741, 742 (1936)

Galliher, E. Wayne

31 Collophane from Miocene brown shales of California. Am As Petroleum G, B 15:257, 269, 10 figs (1931) (*abst*) Rv Geol et Sci conn 12:544 (1932) (*ann*) An Bib Ec G 4:27 (1931) (*abst*) G Zentralbl 45:163 (1931) (*abst*) Eng Index 1931:655 (1931) (*abst*) N Jb 1932 Band 2, Ref 2:113 (1932) Chem Absts 25:2668 (1931)

31a Notes on excrement. Stanford Univ, Micropaleontology B 3:11-13 (1931)

31b Stratigraphic position of the Monterey formation. Stanford Univ, Micropaleontology B 2:71-75 (1931)

32 Geology and physical properties of building stone from Carmel Valley, California. Cal Dp Nat Res, Div Mines, Mining in California, St Mineralogist's Rp 28:14-41 (1932) (*ann*) An Bib Ec G 5:71 (1932)

32a Sediments of Monterey Bay, California. Cal Dp Nat Res, Div Mines, Mining in California, St Mineralogist's Rp 28:42-79, 17 figs (1932) (*abst*) Eng Index 1932:613 (1932)

33 The sulfur-cycle in sediments. Sedimentary Petrology J 3:51-63 (1933)

34 (and **Cushman,** J. A.) Additional new foraminifera from the Miocene of California. Cushman Lab Foram Res, Contr 10: 24-26 (1934) (*abst*) N Jb 1934. Ref 3:624 (1934) Rv Geol et Sci conn 14: 505 (1934)

35 Glauconite genesis. G Soc Am, B 46:1351-1366, 2 pls, 1 fig (1935) (*abst*) G Zentralbl, Abt A, Bd 57:359 (1936)

35a Interstial sedimentation. (*abst*) Pan Am G 63:303 (1935)

35b Geology of glauconite. Am As Petroleum G, B 19:1569-1601 (1935) (*abst*) Rv Geol et Sci conn 16:148 (1936)

Galloway, John

33 (and **Gester,** G. C.)   Geology of Kettleman Hills oil field, California. Am As Petroleum G, B 17:1161-1193 (1933) (*ann*) An Bib Ec G 6:294 (1934) (*abst*) Inst Pet Tech, J 20:2A (1934) Rv Geol et Sci conn 14:74-75 (1934) G Zentralbl 51: 253 (1934)

34 (and **Barbat,** W. F.)   San Joaquin clay, California.   Am As Petroleum G, B 18:476-499 (1934) (*abst*) Inst Pet Tech, J 20:347A-348A (1934) Rv Geol et Sci conn 14:330-331 (1934) (*ann*) An Bib Ec G 7:98 (1935)

35 Accumulation of oil in the Coalinga district. (*abst*) Am As Petroleum G, B 19:1843 (1935) (*ann*) An Bib Ec G 8:357 (1936)

Gannett, Roger W.

32 (and **Ferguson,** H. G.)   Gold quartz veins of the Alleghany district, California. (*abst*) Rv Geol et Sci conn 12:88-90 (1932) U S G S, P P 172:139 pp (1932) (*ann*) An Bib Ec G 5:54 (1932) (*abst*) G Zentralbl, Abt A 48:362-363 (1932) (*abst*) Rv Geol et Sci conn 13:587 (1933) (Review by Knopf, A.)   Ec G 28:399-401 (1933) N Jb 1933, Ref 2:186-187 (1933) (*abst*) Eng Index 1932:632 (1932)

Gardner, E. D.

34 (and **Johnson,** C. H.)   Placer mining in the western United States. U S B M, Inf Circ 6786:1-74... 6787:1-89... 6788:1-81 (1934)

35 (and **Johnson,** F. W.)   Prospecting for lode gold and locating claims of the public domain. U S B M, Inf Circ 6843: 1-18, California 2, 6, 11, 14, 16-18 (1935)

Garner, Kenneth B.

36 Concretions near Mt. Signal, Lower California. Am J Sc 31 (5):301-311 (1936)

Gaylord, E. G.

32 Kettleman Hills development and production problems.   Am Petroleum Inst, Pr Sec 4, Production B 210:40-45 (1932) Oil Weekly, 67 no 11:33-38 (1932)
32a Kettleman Hills field.   Petroleum World (London) 29:364-368 (1932)

Gaylord, H. M.

34 (and **Horton,** F. W.)   Gold, silver, copper, lead, and zinc in California. U S B M, Min Yb 1934:149-158 (1934)... 1935:147-195 (1935)

36 (and **Merrill,** C. W.)   Gold, silver, copper, lead, and zinc in California. U S B M, Min Yb 1936:219-241 (1936)

Gazin, C. L.

31 Geology of the central portion of the Mount Pinos quadrangle, Ventura

Gazin, C. L.—Cont.

and Kern Counties, southern California. (*abst*) G Soc Am, B 42:316 (1931) (*abst*) Pan Am G 54:159 (1930)

31a (**Buwalda,** J. P. and **Sutherland,** J. C.)   Frazier Mountain; a crystalline overthrust slab without roots, west of Tejon Pass, southern California. (*abst*) G Soc Am, B 42:294-295 (1931) N Jb 1933:251 (1933)

33 A Tertiary mammalian fauna from upper Cuyama Drainage Basin, California. (*abst*) Pale Zentralbl 2:324 (1933)

Genth, F. A.

67 Observations on certain doubtful minerals. Ac N Sc Phila, Pr 19:86 (1867)
87 Contributions to mineralogy.   Am Ph Soc, Pr 24: 23-44 (1887).

Gentry, A. W.

33 Ventura County investigation.   Cal Dp Pub Works, Div Water Resources, B 46:35-45 (1933)

Gester, G. C.

33 (and **Galloway,** J.)   Geology of Kettleman Hills oil field, California. Am As Petroleum G, B 17:1161-1193 (1933) (*ann*) An Bib Ec G 6:294 (1934) (*abst*) Inst Pet Tech, J 20:2A (1934) Rv Geol et Sci conn 14:74-75 (1934) G Zentralbl 51:253 (1934) (*abst*) Eng Index 1934:802 (1934)

Geyer, L. E.

36 (and **Metcalf,** R. W.)   Clay. U S B M, Min Yb 1936:867-875, California 869, 871 (1936)

Gianella, Vincent P.

33 Earthquake or landslide? Seism Soc Am, B 23:91-94, 1 pl (1933) (*abst*) Rv Geol et Sci conn 13:620 (1933) Canada, Dom Obs, Pub, Bib Seism 10:344 (1933)

34 (and **Callaghan,** E.)   The earthquake of December 20, 1932, at Cedar Mountain, Nevada and its bearing on the genesis of Basin Range structure.   J G 42:1-22 (1934) (*Review*) Sc Progress 29:501 (1935) (*abst*) G Zentralbl 51:455 (1934)
34a New features of the geology of the Comstock lode. M Metal 15:298-300 (1934)

Gilbert, G. K.

31 Studies of Basin Range structure. (*abst*) G Zentralbl 143:97-98 (1930-1931)

Gilbert, J. C.

33 (**Collins,** L. B.; and **Siemon,** J. H.) Marine oil drilling in California. Petroleum World (London) 30:327-329 (1933) Petroleum World 1933 (October):15-18 (1933)

Goldsmith, E.

76 On sonomaite. Ac N Sc Phila, Pr 28:263-264 (1876)

Gonyer, F. A.

32 (and **Irving**, J.; **Vonson**, M.) Pumpellyite from California. Am Mineralogist 17:338-342 (1932) (*abst*) Rv Geol et Sci conn 13:84 (1933) Min Absts 5:233 (1933)

34 (and **Moehlman**, R. S.) Monticellite from Crestmore, California. Am Mineralogist 19:474-476 (1934) (*abst*) N Jb 1935, Ref 1:62 (1935) Min Absts 6:96 (1935)

Goodrich, H. B.

32 Early discoveries of petroleum in the United States. Ec G 27:160-168 (1932)

Goodspeed, George E.

32 Micro-structures of gold quartz-veins. (*abst*) Pan Am G 58:79 (1932)

Goold, W. D.

30 (**Jensen**, J.; **McDowell**, G.; and **Gwin**, M. L.) Deep sand development at Santa Fe Springs. *In* Petroleum Div and Tech. Am I M Eng, Tr 1930:310-321 (1930)

Gore, F. D.

27 Oil shale in Santa Barbara County, California. (*abst*) N Jb 1927, Ref 2; Abt B:114 (1927)

Goudey, Hatfield

36 Minerals—Ritter Range, California. The Mineralogist 4, no 5:7-8, 26, 28-29 (1936)

Goudkoff, Paul P.

31 Age of producing horizon at Kettleman Hills, California. Am As Petroleum G, B 15:839-842 (1931) (*ann*) An Bib Ec G 4:317 (1931) (*ann*) An Bib Ec G 4:317 (1931) (*abst*) G Zentralbl 45:389-390 (1931) (*abst*) N Jb 1932, Band 2, Ref 2:113 (1932)

34 Subsurface stratigraphy of Kettleman Hills oil field, California. Am As Petroleum G, B 18:435-475 (1934) (*ann*) An Bib Ec G 7:100 (1935) (*abst*) Rv Geol et Sci conn 14:276 (1934) Inst Pet Tech, J 20:348A (1934)

Gould, Charles N. (Chairman)

33 Committee on stratigraphic nomenclature. Classification and nomenclature of rock units.   G Soc Am, B 44:423-459 (1933)

Gould, H. W.

31 Quicksilver. Min Ind 39:529-534, California 532 (1931)... 40:481-487, California 485 (1932)

Grant, Ulysses S. IV

31 (and **Gale**, H. R.) Catalogue of the marine Pliocene and Pleistocene mollusca of California. San Diego Soc N H, Mem 1: 1036 pp, 15 figs, 32 pls (1931) (*abst*) G Zentralbl Abt A 47:393-395 (1932) N Jb 1933, Ref 3:168-174 (1933) Nature 129: 135 (1932) Pale Zentralbl 2:186-190 (1933) Biol Absts 8:242-243, entry 2118 (1934)

31a (and **Oldroyd**, I. S.) A Pleistocene molluscan fauna from near Goleta, Santa Barbara County, California. Nautilus 44: 91-94 (1931) (*abst*) N Jb 1934, Ref 3:283 (1934) G Zentralbl 46:425 (1932) Pale Zentralbl 1:44 (1932)

32 (and **Soper**, E. K.) Geology and paleontology of a portion of Los Angeles, California. G Soc Am, B 43:1041-1067 (1932) (*amn*) An Bib Ec G 5:341 (1932) (*abst*) Pan Am G 57:370-371 (1932) G Soc Am, B 44:148 (1933) Pale Zentralbl 4:162 (1934)... 5:91-92 (1934) Biol Absts 9:1857, entry 16826 (1935)

33 Notes on Searlesia. (*abst*) G Soc Am, B 44:220 (1933) Pale Zentralbl 4:276 (1934)

34 (and **Quayle**, E. H.) A new middle Miocene *Neptunea* from California. Nautilus 47:91-93 (1934)

34a (and **Strong**, A. M.) Fossil mollusks from the vertebrate-bearing asphalt deposits at Carpinteria, California. S Cal Ac Sc, B 33:7-11 (1934) (*abst*) Pan Am G 59:375 (1933) G Soc Am, Pr 1933 1:390 (1934) (*abst*) Rv Geol et Sci conn 14:244 (1934)

34b (and **Soper**, E. K.) Stratigraphy of western Santa Monica Mountains. (*abst*) Pan Am G 61:308 (1934) (*abst*) G Soc Am, Pr 1934:310-311 (1935)

34c (and **Strong**, A. M.) Pliocene and Pleistocene mollusca of Santa Barbara. (*abst*) Pan Am G 62:71-72 (1934) (*abst*) G Soc Am, Pr 1934:386-387 (1935)

35 Summary of marine Pleistocene of California. (*abst*) Pan Am G 64:73-74 (1935) (*abst*) G Soc Am, Pr 1935:349 (1936)

35a (and **Putman**, W. C.) Barrancos and arroyos in California. (*abst*) Pan Am G 63:307-308 (1935) (*abst*) G Soc Am 1935:331 (1935)

Graves, McDowell

30 (**Jensen**, J.; **Goold**, W. D.; and **Gwin**, M. L.) Deep sand development at Santa Fe Springs. *In* Petroleum Dev and Tech Am I M Eng, Tr 1930:310-321 (1930)

Green, Norman B.

33 Reinforced concrete in the Long Beach earthquake. Eng News-Rec 110: 560-562, 2 figs (1933) (*abst*) Canada, Dom Obs, Pub, Bib Seism 10:309 (1933)

Greger, Darling K.

34 Notes on a collection of *Dentalium neohexagonum* S. and P. G Mag 71:236-237 (1934)

Gregory, Herbert E.

25 Reconnaissance traverse from Mojave Ulojane, California, to Rock Creek, Utah. (*abst*) N Jb Band 1:257-258 (1925)

Gregory, J. W.

31 The geological history of the Pacific Ocean. Q J G Soc London 86:lxxii-cxxxvi (1930) (*abst*) N Jb 1931, Ref 3:237-240 (1931)

Greig, T. W.

32 Temperature of formation of the ilmenite of the Engels copper deposit—a discussion. Ec G 27:25-38 (1932)

Grieger, J. M.

34 San Diego County, California, gem mines not exhausted. Oregon Mineralogist 2, no 10:7-8, 20 (1934)

Griffith, Lloyd

29 Preparedness of the oil companies for a major disaster in the Los Angeles basin. Seism Soc Am, B 19:156-161 (1929) (*abst*) Canada, Dom Obs, Pub, Bib Seism 10:72 (1930)

Grinnell, Joseph

32 A relic shrew from central California. (A relic of a palustrine fauna) Cal Univ, Pub Zool 38:389-390 (1932) Biol Absts 7:1778, entry 17633 (1933)

Gross, Paul L. K.

35 (and Eckis, R.) Porosity and sorting of California fanglomerates. (*abst*) Pan Am G 64:77 (1935) (*abst*) G Soc Am, Pr 1935:353 (1936)

Grover, N. C.

31 (McGlashan, H. D.; Henshaw, F. F.; and Canfield, G. H.) Surface water supply of the United States 1928 Part XI, Pacific Slope basins in California. U S G S, W-S P 671: IX, 304 pp (1931)... 1929, 691:IX, 294 pp (1931)... 1930, 706:IX, 317 pp. (1932)... 1931, 721:XI, 497 pp (1932)... 1931, 721:XI, 497 pp (1932)... 1932, 736:XI, 415 pp (1933)... 1933, 751:XI, 376 pp (1935)

32 Water resources branch, California. U S G S, An Rp 1932:57 (1932)

Grunsky, C. E.

31 Comments on a few dams and reservoirs. Military Eng 23, no 128:135-139 (1931)

Gudger, E. W.

33 The opah or moonfish, Lampris Luna, on the coasts of California and Hawaii. (Attention is called to a fossil from the Lompoc beds in California.) Am Nat 65:531-540, 1 fig (1931) Biol Absts 7:1001, entry 9889 (1933)

Guiberson, W. R.

34 The challenge of Coalinga. Petroleum World An Rv 1934:173-177 (1934)

Gunter, Herman

31 Fuller's earth. Min Ind 40:608-710, California 609 (1932)

Gutenberg, Beno.

31 Microseisms in North America. Seism Soc Am, B 21: 1-24, 4 figs (1931) (*abst*) G Zentralbl, Abt A 47:231 (1932)

31a (and Richter, C. F.) On supposed discontinuities in the mantle of the earth. Seism Soc Am, B 21:216-223 (1931)

32 (Wood, H. O., and Buwalda, J. P.) Experiments testing seismographic methods for determining crustal structure. (*abst*) Pan Am G 58:65-66 (1932) Seism Soc Am, B 22:185-242 (1932)

32a (Richter, C. F., and Wood, H. O.) The earthquake in Santa Monica Bay, California, on August 30, 1930. Seism Soc Am, B 22:138-154, 1 fig, 2 pls (1932) (*abst*) N Jb 1933, Ref 2:573-574 (1933)

32b Travel time curves at small distances and wave velocities in southern California. Beitr Geoph 35 heft 1:6-45, 11 figs (1932) (*abst*) Canada, Dom Obs, Pub, Bib Seism 10:229 (1932)

33 Tilting due to glacial melting. J G 41:449-467, California 464-465 (1933)

34 The propagation of the longitudinal waves produced by the Long Beach earthquake. Gerl Beitr 41:114-120, 3 figs (1934) (*abst*) N Jb 1934, Ref 2:545 (1934)

35 (and Buwalda, J. P.) Seismic reflection profile across Los Angeles Basin. (*abst*) Pan Am G 63:303 (1935) (*abst*) G Soc Am, Pr 1935:327 (1936)

35a (and Buwalda, J. P.) Investigation of overthrust faults by seismic methods. Science n s 81, no 2103:384-386 (1935)

Gutenberg, G.

34 Das "Seismological Laboratory" in Pasadena (the seismological laboratory in Pasadena) Ergebnisse der Kosmischen Physik 2:123-237 Leipzig (1934) (*ann*) An Bib Ec G 8:198 (1935)

Gwin, M. L.

30 (Jensen, J.; McDowell, G.; and Goold, W. D.) Deep sand development at Santa Fe Springs. *In* Petroleum Dev

Gwin, M. L.—Cont.

and Tech. Am I M Eng, Tr 1930:310-321 (1930)

Hake, B. F.

32 Scarps of the southwestern Sierra Nevada, California. (*abst*) Zs Geom 7: 53-54 (1932)

Hall, C. L.

97 The gravel fields of northern California. M Sc Press 74:113 (1897)

Hall, E. R.

30 A bassarisk and a new mustelid from the later Tertiary of California. J Mammalogy 11:23-26, 2 figs (1930) (*abst*) Pale Zentralbl 2:66 (1932)

33 (and **Stock**, C.) The Asiatic genus Eomellivora in the Pliocene of California. J Mammalogy 14:63-65, 1 pl (1933) Biol Absts 8:834, entry 7628 (1934)

34 Rodents and lagomorphs from the Barstow Beds of southern California. Biol Absts 8:281, entry 2545 (1934)

Hammar, Harold E.

33 (and **Trask**, P. D.) Source beds in Mesozoic rocks west of Sacramento River, California. (*abst*) Pan Am G 59:229 (1933)

33a (and **Trask**, P. D.) Some relations of the organic constituents of sediments to the formation of petroleum. (*abst*) Wash Ac Sc, J 23:568 (1933)

34 (and **Trask**, P. D.) Preliminary study of source beds in late Mesozoic rocks on west side of Sacramento Valley, California. Am As Petroleum G, B 18: 1346-1373 (1934) (*abst*) Wash Ac Sc, J 24: 491-492 (1934) (*abst*) N Jb 1935, Ref 2:688 (1935) (*ann*) An Bib Ec G 8:154 (1936) (*ann*) An Bib Ec G 7:312 (1934) (*abst*) Rv Geol et Sci conn 15:207-208 (1935)

35 (and **Trask**, P. D.) Organic content of sediments. (*abst*) Wash Ac Sc 25:508 (1935) Am Petroleum Inst, Productions, B 214 (1934) Oil Gas J 33, no 27, 28, 29: 43-46, 40-41, 36-39 (1934) Petroleum World (London) 32:19-20 (1935)

Hampton, Edgar Lloyd

33 The California earthquake by engineer and architect. Tech Eng New 14, no 5:94-95, 104 (1933) (*abst*) Canada, Dom Obs, Pub. Bib Seism 10:345 (1933)

Hanks, Henry Garber

82 Gold nuggets. Cal St Mineralogist's Rp 2:147-150 (1882)

Hanna, G. Dallas

30 Porosity of diatomite. Eng M J 130: 7-8, 7 figs (1930)

Hanna, G. Dallas—Cont.

31 Diatoms and silicoflagellates of the Kreyenhagen shale. Cal Dp Nat Res, Div Mines, Mining in California, St Mineralogist's Rp 27:187-201, 5 pls (1931) (*abst*) Pale Zentralbl 1: 28 (1932) (*abst*) G Soc Am, B 42:306 (1931)

31a Illustrating fossils. J Paleontology 5:49-68, 1 pl, 2 figs (1931)

31b Geology of Sharktooth Hill, Kern County, California. (*abst*) G Zentralbl 43:270-271 (1930-1931) N Jb 1933, Ref 3: 359-360 (1933) Biol Absts 5:25386 (1931)

32 The Monterey shale of California at its type locality with a summary of its fauna and flora. Biol Absts 6:2181 (1931) (*abst*) N Jb 1930, Ref 2:239 (1930)

32a *Desmostylus* tooth dredged in Monterey Bay. (*abst*) G Soc Am, B 43:291 (1932) (*abst*) N Jb 1934, Ref 3:159 (1934)

32b The diatoms of Sharktooth Hill, Kern County, California. Cal Ac Sc, Pr (4) 20:161-263, 17 pls (1932)

33 The name "Lillis Formation" in California geology. Am As Petroleum G, B 17:81-84 (1933)

33a (and **Smith**, A. G.) Two new species of *Monadenia* from northern California. Nautilus 46, no 3:79-86, 2 pls (1933) Biol Absts 8:243, entry 2121 (1934)

34 Land shells from the upper Eocene Sespe deposits, California. Wash Ac Sc, J 24:539-541, 3 pls (1934)

34a Additional notes on diatoms from the Cretaceous of California. J Paleontology 8:352-355, 1 pl (1934) (*abst*) G Soc Am, Pr 1933 1:377 (1934) (*abst*) Rv Geol et Sci conn 15:216-217 (1935)

35 (**Taff**, J. A.; and **Cross**, C. M.) Chico Cretacic at type-locality. (*abst*) Pan Am G 64:72 (1935) (*abst*) G Soc Am, Pr 1935:348 (1936)

35a Interesting whale jaw from Kern County. (*abst*) Pan Am G 64:79-80 (1935) (*abst*) G Soc Am, Pr 1935:419 (1936)

35b (and **Cushman**, J. A.) Foraminifera from the Eocene near Coalinga, California. Biol Absts 9:457, entry 3894 (1935)

36 A new land shell from the Eocene of California, J Paleontology 10, no 5:416-417, 1 text fig (1936)

36a (and **Vonson**, M.) Borax Lake, California. Cal Dp Nat Res, Div Mines, M G, J, St Mineralogist's Rp 32:99-108, 5 text figs (1936).

Harden, B. D.

35 The Spanish mine: brief history and recent metallurgy. M Metal 16:415-417 (1935) (*abst*) Eng Index 1935:523 (1935)

Harding, C. R.

31 Location and design of the Southern

Harding, C. R.—Cont.

Pacific Company's Suisun Bay Bridge as affected by consideration of earthquakes. (*ann*) An Bib Ec G 4:112 (1931)

Hardy, Charles

31 Barium and strontium. Min Ind 39:64-68, California 65, 66 (1931)... 40:63-67, California 64 (1932)... 41:53-57, California 54 (1933)... 42:57-61, California 59 (1935)

Hartmann, Miner L.

31 California's diatoms supply a wide and varied market. Chem & Met Eng 38:652-653 (1931) (*ann*) An Bib Ec G 4:273 (1931)

Haselwood, F. W.

35 Grizzly Dome drops 75,000 yards of granite into Feather River Canyon. Cal Dp Pub Works, Cal Highways April: 4, 28 (1935)

Hatai, Kotora

36 A note on the distribution of certain species of marine invertebrates. Biogeographical Soc Japan, B 6, no 12:123-141 (1936)

Hatmaker, Paul

30 (and **Davis,** A. E.) Abrasive materials. U S B M, Min Res, Calendar Year 1930:151-169, California 152, 161, 163, 164 (1932)... 1931:111-130, California 112, 121, 124 (1933)

31 (and **Bowles,** O.) Trends in the production and uses of granite as dimension stone. U S B M, Rp Invest 3065:21 pp, 11 figs (1931) (*abst*) U S B M, List Pub 1910-1932:115 (1933)

31a (and **Middleton,** J.) Fuller's earth. U S B M, Min Res, Calendar Year 1931: 73-98, California 84, 91 (1933)

31b Diatomite. U S B M, Inf Circ 6391:20 pp, California 4, 17 (1931)

32 (and **Coons,** A. T.) Lime. U S B M, Min Yb 1932-1933:629-637, California 631, 632 (1933)... 1934:861-871, California 863 (1934)

32a Pumice and pumicite. U S B M, Inf Circ 6560:1-24, California 11, 13, 14, 21, 22 (1932)

Hay, Oliver P.

30 On the fossil mammalia of the first interglacial stage of the Pleistocene of the United States. Wash Ac Sc, J 20: 501-509 (1930)

Hazzard, John C.

33 Notes on the Cambrian rocks of the eastern Mojave Desert, California. Cal Univ, Dp G, B 23:57-80, 1 pl, 1 fig (1933)

Hazzard, John C.—Cont.

36 (and **Mason,** J. F.) Stratigraphy and lithology of the Middle Cambrian formations of the Providence and Marble Mountains, San Bernardino County, California. (*abst*) G Soc Am, Pr 1935:408-409 (1936) (*abst*) Pan Am G 63:369-370 (1935)

36a (and **Mason,** J. F.) Middle Cambrian formations of the Providence and Marble Mountains, California. G Soc Am, B 47:229-240, 1 fig (1936)

Head, E. R.

35 New light oil field discovery made in Montebello area. Cal Oil World 27, no 28:4 (1935)

Heald, K. C.

30 Determination of geothermal gradients in oil fields on anticlinal structure. Am Petroleum Inst, Pr 11, no 1:102-109 (1930) (*abst*) Eng Index 1931:660 (1931)

Heck, N. H.

29 Earthquake investigation in the United States. U S Coast S, Serial no 456:21 pp (1929)

30 Earthquakes—a challenge to science. Sc Mo 31:113-125, California 116-117 (1930)

31 Doing something about earthquakes. Sc Mo 32:365-367, California 366, 367 (1932)

31a Seismology and engineering. The Military Eng 1931:131-134, California 132, 134, 6 illus (1931)

31b Coming to grips with the earthquake problem. Smiths An Rp 1931:361-380, California 365-366, 370-371, 377, 378, 379, pls, figs, maps (1931) Franklin Inst, J 212:269-303, California 270, 278, 285, 289, 297, 299, 301, figs, maps (1931)

32 Structural hazard of earthquakes measured by new instruments. Eng News-Rec 109:288-289 (1932)

33 (and **Neumann,** F.) Destructive earthquake motions measured for first time. Eng News-Rec 110:804-807, 8 figs, 1 table (1933) (*abst*) Canada, Dom Obs, Pub, Bib Seism, 10:345 (1933)

33a Strong-motion records of Long Beach earthquake. Eng News-Rec 110: 442-443 (1933)

35 A review of outstanding problems in strong-motion vibration work. Seism Soc Am, B 25:343-349 (1935)

35a Investigation of strong earthquake motions in California. Wash Ac Sc, J 25: 513 (1935)

35b (**Wood,** H. O.; and **Allen,** M. W.) Destructive and near-destructive earthquakes in California and western Nevada, 1769-1933. U S G S, Sp Pub 191:24 pp (1934)

Heck, N. H.—Cont.

35c A new map of earthquake distribution. Geog Rv 25:125-131, California 128, 130, map (1935)

36 Observations on recent progress in seismology. (abst) Pan Am G 65:157 (1936)

Hedges, J. H.

32 Potash. U S B M, Min Yb 1932-1933:763-775, California 767 (1933)... 1934: 1031-1046, California 1036 (1934)... 1935: 1137-1164, California 1143, 1154 (1935)... 1936:1007-1021, California 1012 (1936)

Heikes, V. C.

31 California's metal mines output. Mining Rv 33, no 1:13-14 (1931) (abst) Eng Index 1931:869 (1931)

32 Gold, silver, copper, lead, and zinc in California. U S B M, Min Res Calendar Year 1929:431-470 (1932)... 1930:963-1027 (1933) (abst) Eng Index 1931:869 (1931)... 1932:803 (1932)

32a (and Merrill, C. W.) Gold, silver, copper, lead, and zinc in California. U S B M, Min Yb, Stat App 1932-1933:199-215 (1934)

33 Gold, silver, copper, lead and zinc in California, Nevada, and Oregon. U S B M, Min Yb 1932-1933:121-136, California 121-128 (1933)

Heller, A. H.

31 Possible new productions in Kettleman area. Oil B 17:434-440 (1931) (abst) Eng Index 1931:1036 (1931)

31a Valuation of oil properties. Oil B 17:269-271, 363-365, 404 (1931)

Henderson, Charles W.

32 Gold and silver. U S B M, Min Yb 1932-1933:11-26, California 14, 18, 19, 21-22 (1933)... 1934:25-52, California 40, 41. 47, 49 (1934)... 1935:27-44, California 27, 28, 30, 35-36, 40, 41 (1935)

33 The history and influencing of mining in the western United States. In Ore deposits of the western states. Am I M Eng, Pub:730-784, California 733 (1933)

36 Gold and silver. U S B M, Min Yb 1936:91-105, California 91, 92, 93, 95, 101, 103 (1936)

36a (and Dunlop, J. P.) Gold and silver. U S B M, Min Yb 1936:91-105, California 92, 93, 95, 101, 103, (1936)

Hendy, N. Ingram

33 A preliminary note on the distribution of marine diatoms during the Tertiary period. J Botany 71:111-118, 1 fig (1933)

Henny, Gerard

30 McLure shale of the Coalinga region,

Henny, Gerard—Cont.

Fresno and Kings Counties, California. Am As Petroleum G, B 14, no 4:403 (1930)

31 Presence of the McLure shale on west side of the San Joaquin Valley. (ann) An Bib Ec G 4:100 (1931) G Zentralbl, Abt A 48:427 (1933) Petroleum World 1930 (August):97-99, 117 (1930)

Henshaw, F. F.

31 (Grover, N. C.; McGlashan, H. D.; and Canfield, G. H.) Surface water supply of the United States, 1928 Part XI Pacific Slope basins in California. U S G S, W-S P 671:IX, 304 pp (1931)... 1929, 691:IX, 294 pp (1931)... 1930, 706 IX, 317 pp (1932)... 1931, 721:XI, 427 pp (1932)... 932, 736:XI, 415 pp (1933)

Herlihy, K. V.

32 (and Middleton, J.) Clay. U S B M, Min Res, Calendar Year 1930:171-180, California 174, 176, 178 (1932)

33 (and Kiessling, O. E.) Clay. U S B M, Min Res, Calendar Year 1931:237-242, California 239 (1933)

33a (and Kudlich, R. H.) Clay. U S B M, Min Yb 1932-1933:639-645, California 641 (1933)

Herold, C. Lathrop

34 Fossil markings in the Carmelo Series. (Upper Cretaceous) Point Lobos, California. J G 42:630-640 (1934) (abst) Rv Geol et Sci conn 16:79-80 (1936)

35 Geology of Salinas quadrangle. (abst) Pan Am G 63:315-316 (1935) (abst) G Soc Am, Pr 1935:337 (1936)

36 Distribution of Eocene rocks in Santa Lucia Mountains California. Am As Petroleum G, B 20:491-494 (1936)

Herold, Stanley C.

30 Problem of drainage at Kettleman Hills. Oil Gas J 28, no 45:40, 146-147 (1930) Oil Weekly 57, no 2:32-38 (1930)

31 On and off the dome at Kettleman Hills, California. Oil Weekly 61, no 3:36, 38, 42, 65, 4 figs (1931) Oil Gas J 29:100, 165, 166, 169, 4 figs (1931) Oil B 17, no 3:188-191, 238 (1931) (ann) An Bib Ec G 4:100 (1931) (abst) Inst Pet Tech, J 17: 344A (1931) (abst) Eng Index 1931:1036 (1931)

33 How much will an oil field produce? Petroleum World 1933 (June):23-27 (1933)

34 Geology and technology go forward. Petroleum World 1934:32-44 (1934)

35 (and Hoots, H. W.) Natural gas resources of California, geology of natural gas. Am As Petroleum G 1935:113-220, 36 figs (1935) (abst) Rv Geol et Sci conn 16, fasc 10:579-580 (1936)

Herring, C. T.

36 (Davis, H. W.; and Kiessling, O. E.) Iron ore, pig iron, ferro-alloys and steel. U S B M, Min Yb 1936:365-394, California 375 (1936)

Hershey, O. H.

05 The Quaternary of southern California. (abst) N Jb 1905, Band 145-146 (1905)

Hertel, F. W.

30 Ventura Avenue oil field, Ventura County, California. (abst) N Jb 1930, Ref 2:241 (1930)

31 (and Jensen, J.) Development in a part of the Ventura Avenue oil field. In Petroleum Dev and Tech, Am I M Eng, Tr 1931:149-156 (1931) Oil Weekly 59, no 6:34-43 (1930)

Hertlein, Leo George

31 Preliminary report on the paleontology of the Channel Islands, California. (abst) G Zentralbl 43:269-270 (1930-1931) Biol Absts 5:950, entry 9481 (1931)

31a Changes of nomenclature of some recent and fossil Pectinidae from Japan, Porto Rico, South America, New Zealand, and California. J Paleontology 5:367-369 (1931) (abst) Pale Zentralbl 3:425 (1933)

33 Additions to the Pliocene fauna of Turtle Bay, Lower California, with a note on the Miocene diatomite. J Paleontology 7:439-441 (1933) Biol Absts 9:1616, entry 14669 (1935)

33a A new Gryphaeoid oyster from the Eocene of California. San Diego Soc N H, Tr 7:277-280 (1933) (abst) N Jb 1933, Ref 3:1255 (1933) Pale Zentralbl 5: 141 (1934)

33b Some of the common rocks of the San Francisco Bay region. The Aquarium J 6:106-107 (1933)

34 New oysters and a new pecten from the Tertiary of California. S Cal Ac Sc, B 33:1-6, pl 1-2 (1934) (abst) Rv Geol et Sci conn 14:244 (1934)

34a Three new specific names for west American fossil mollusca. Biol Absts 8:2365, entry 21109 (1934)

34b A new Pecten from the San Diego Pliocene. (abst) G Zentralbl 5:139 (1934)

Hess, C. J.

34 Another record of the fossil vole Mimmomys primus (Wilson) from California. J Mammalogy 15:246 (1934) (abst) Rv Geol et Sci conn 15:617 (1935)

Hess, Frank L.

33 Tungsten. U S B M, Min Res, Calendar Year 1930:179-207, California 179, 182 (1933)... 1931:43-50, California 43, 44

Hess, Frank L.—Cont.

(1934)... 1932-1933:Min Yb:271-279, California 275 (1933)... 1934:435-444, California 339-340 (1934)... 1935:491-498, California 494 (1935)

33a The pegmatites of the western states. Am I M Eng, Lindgren Volume: 526-536 (1933)

34 Molybdenum. U S B M, Min Yb 1934:417-434, California 426-427 (1934)

Hester, James

31 (Byerly, P., Marshall, K.) The natural periods of vibration of some tall buildings in San Francisco. Seism Soc Am, B 21:268-276, 3 figs (1931) (abst) Canada, Dom Obs, Pub, Bib Seism 10:228 (1932)

Hewett, D. F.

33 Sedimentary manganese deposits. In Ore deposits of the western states. Am I M Eng:488-491, California 490 (1933) (abst) N Jb 1934, Ref 2:722 (1934)

33a (and Pardee, J. T.) Manganese in western hydrothermal ore deposits. In Ore deposits of the western states. Am I M Eng, Lindgren Volume:671-682, California 671, 672, 680 (1933)

36 (and others) Mineral resources of the region around Boulder Dam. U S G S, B 871:197 pp (1936)

Hicks, N. H.

29 (and Bodle, R. R.) United States earthquakes. U S, Coast S 1928 Serial 483:31 pp, 5 figs (1930)... 1929 Serial 511: 61 pp, California 11-15, 1 fig (1931)... 1930 Serial 539:25 pp, California 3-14, 2 figs (1932)... 1931 Serial 553:26 pp, California 14-20 (1932)... 1932 Serial 563:21 pp, California 6-15 (1934)

Hidden, W. E.

91 Mineralogical notes. Am J Sc (3) 41:438 (1891)

Hight, William

33 Graphic history, development and production of California oil fields. Cal Oil World 25, no 29:8-9 #1 Inglewood field... 25, no 31:8, 10 #2 Rosecrans field... 25, no 33:8, 10 #3 Torrance field... 25, no 35:12, 16 #4 Playa del Rey field... 25, no 37:8, 10 #4 Mount Poso field... 25, no 39:8 #6 Seal Beach field... 25, no 41:8 #7 Alamitos Heights field... 25, no 43:8, 15 #8 Elk Hills field... 25, no 45:8 #9 Richfield field... 25, no 47:8 #10 Montebello field... 25, no 49:8 #11 Kern Front field... 25, no 51:8-9 #12 Elwood field... 26, no 1:8, 12 #13 Kettleman Hills field... 26, no 3:8, 9 #14 Ventura field... 26, no 5:8, 10 #15 Dominguez field... 26, no 7:8-9 #16 Long Beach field (1933)

Hill, H. R.

34 Notes on bentonite. Pacific Mineralogist. 1, no 1:12-13 (1934)

Hill, Mason Lowell

31 Structure of the San Gabriel Mountains north of Los Angeles, California. (abst) G Zentralbl 45:495-596 (1931) N Jb 1933, Ref 3:577-579 (1933)

32 Mechanics of faulting near Santa Barbara. J G 40:535-556, 9 figs (1932) (abst) G Soc Am, B 43:149-150 (1932) (abst) Pan Am G 57:72 (1932) (abst) Rv Geol et Sci conn 13:557-558 (1933) (abst) Eng Index 1932:612 (1932)

33 Diagnostic phenomena for direction of fault-slip as illustrated in Kettleman Hills. (abst) Pan Am G 59:319 (1933) (abst) G Soc Am, Pr 1933 1:317 (1934)

34 Origin of faulting in Kettleman Hills. (abst) Pan Am G 61:317-318 (1934) (abst) G Soc Am, Pr 1934:322, 323 (1935)

Hill, R. A.

32 New emergency water supply for Santa Barbara. Western City 7, no 2:9-13 (1932)

Hillis, Donnil

33 Cracks produced by Long Beach California earthquake. Am As Petroleum G, B 17:739-740 (1933)

Hinderlider, M. C.

31 (with Savage, J. L.; Berkey, C. P.; Louderback, G. D.; and Williams, I. A.) Report of consulting board on safety of the proposed Pine Canyon dam, Los Angeles County. Cal Dp Pub Works, Div Water Resources May 1931:22 pp, 8 pls, 1 map (1931)

Hinds, Norman E. A.

31 Most ancient formation in the Klamath Mountains. (abst) G Soc Am, B 42: 292-293 (1931)

32 Paleozoic eruptive rocks of the southern Klamath Mountains, California. Cal Univ Dp G Sc, B 20:375-410, 2 figs, 2 pls (1932) (abst) Zs Vulkan 15:145 (1933) (abst) Eng Index 1932:613 (1932)

32a Diastrophic epochs in the southern Klamath Mountains, California. (abst) Pan Am G 57:153-154 (1934) (abst) G Soc Am, B 43:273-274 (1932)

33 Geologic formations of the Redding-Weaverville districts, northern California. Cal Dp Nat Res, Div Mines, M G, J, St Mineralogist's Rp 29:77-122 (1933) (ann) An Bib Ec G 6:222 (1934) (abst) G Soc Am, Pr 1934:315 (1935) (abst) Eng Index 1934:516 (1934) (abst) G Zentralbl, Abt A, Bd 53:330 (1934)

Hinds, Norman E. A.—Cont.

34 Late Cenozoic history of southern Klamath Mountains. (abst) Pan Am G 61:315-316 (1934)

34a Geology of the Weaverville district, northern California. (abst) Pan Am G 61:312 (1934)

34b The Jurassic age of the last granitoid intrusives in the Klamath Mountains and Sierra Nevada, California. Am J Sc 27:287-294 (1934) (ann) An Bib Ec G 7:20 (1935) (abst) Zs Vulkan 16:78 (1934)

35 Mesozoic and Cenozoic eruptive rocks of the southern Klamath Mountains, California. Cal Univ, Dp G, B 23:313-380, 16 figs, 1 map (1935) (ann) An Bib Ec G 8:255 (1936)

Hinton, Martin A. C.

32 Note on "Cosomys" Wilson, from the Pliocene of California. J Mamalogy 13:280-281 (1932)

Hirschkind, W.

31 Alkaline lake brines supply western soda producers. Chem Met Eng 38:657-659 (1931) (abst) Eng Index 1931:214 (1931)

Hobson, Henry D.

32 The stratigraphic significance of foraminifera from the type San Lorenzo formation, California. Stanford Univ, Micropaleontology, B 3:30-40 (1932) (abst) Pan Am G 58:148 (1932) (abst) G Soc Am, B 44:217 (1933) (abst) Pale Zentralbl 4:148 (1934)

35 (and Cushman, J. A.) A foraminiferal faunule from the type San Lorenzo formation, Santa Cruz County, California. Cushman Lab Foram Res, Contr 11:65-73, pls, figs (1935) (abst) Rv Geol et Sci conn 15:606 (1935)

Hodge, Edwin T.

29 Earthquake insurance risks of the Pacific Coast. 55th Annual Meeting Fire Underwriters Association of the Pacific Coast, February 1929 (abst) Rv Geol et Sci conn 11:415-416 (1930)

Hodges, F. C.

32 (and Johnson, A. M.) Subsurface storage of oil and gas in the Brea-Olinda and Lompoc fields. Cal Dp Nat Res, Div Oil and Gas, Cal Oil Fields 17, no 4:5-12 (1932) (abst) Rv Geol et Sci conn 14:73-74 (1934)

Hoffman, Thelma

35 Natural gas and natural gasoline. Petroleum World (London) 32:156-157 (1935)

Hollick, Arthur

32 Descriptions of new species of Tertiary cycads, with a review of those previously recorded. Torrey Bot Club B 59; no 4:169-189, California 172, 14 pls (1932) Biol Absts 7:180, entry 1803 (1933)

Hollister, J. S.

35 (and Reed, R. D.) Paleogeology of southern California. (abst) Am As Petroleum G, B 19:1841 (1935) (ann) An Bib Ec G 8:358 (1936)

35a (and Reed, R. D.) Geology of the Transverse Ranges. (abst) Am As Petroleum G, B 19:136 (1935)

36 (Reed, F. D.; and Ashauer, H.) Sedimentation und faltung im sudlichen Kalifornien. Stille-Festschrift 1936:232-233 (1936)

36a (and Reed, R. D.) Structural evolution of southern California. Am As Petroleum G, B 20:1529-1692, 57 figs, 9 pls, 6 tables, map (1936)

Holmes, Arthur

35 When will Lassen Peak again erupt? Sc Mo 40:21-33 (1935)

Holway, Ruliff S.

06 Eclogites in California. (abst) N Jb 106 Band 2:218 (1906)

16 Preliminary report on the recent volcanic activity of Lassen Peak. (abst) Zs Vulkan 2:243-246 (1915-1916)

Hoots, Harold W.

30 Oil shale in a producing oil field in California. (abst) N Jb 1930, Ref 2:239-240 (1930)

31 Geology and oil resources along the southern border of the San Joaquin Valley. (abst) G Zentralbl 43:374-375 (1930-1931) N Jb 1931, Ref 2:541-542 (1931) (abst) Eng Index 1930:1300 (1930) Zs Prak G 39:62-63 (1931) (abst) Zs Geom 7:61-62 (1932)

31a Geology of eastern part of the Santa Monica Mountains, California. U S G S, P P 165:83-134 (1931) (ann) An Bib Ec G 4:100 (1931) (abst) G Zentralbl, Abt A, 47:187-188 (1932) (abst) Eng Index 1931:1036 (1931)

32 Excursion in Los Angeles basin and Santa Monica Mountains. Int G Cong, Guide Book 15:43-48 (1932)

32a General geology of the eastern part of the Santa Monica Mountains. Int G Cong, Guide Book 15:40-43 (1932) (abst) N Jb 1936, Ref 3:60-61 (1936)

32b General geology of the Los Angeles basin. Int G Cong, Guide Book 15:23-26 (1932) (abst) N Jb 1936, Ref 3:57-58 (1936)

32c Oil development in the Los Angeles basin. Int G Cong, Guide Book 15:

Hoots, Harold W.—Cont.

26-30 (1932) (abst) N Jb 1936, Ref 3:58-59 (1936)

35 (Blount, A. L.; and Jones, P. H.) Marine oil shale, source of oil in Playa del Rey field, California. Am As Petroleum G, B 19:172-206 (1935) (ann) An Bib Ec G 8:142 (1936) (abst) Rv Geol et Sci conn 15:209 (1935)

35a (and Herold, S. C.) Natural gas resources of California. Geology of natural gas. Am As Petroleum G 1935:113-220, 36 figs (1935) (abst) Rv Geol et Sci conn 16, fasc 10:579-580 (1936)

36 Migration of oil in California. Am As Petroleum G, B 20:613-615 (1936)

36a Recent discoveries and present oil supply in California. Am As Petroleum G, B 20:939-950, 4 figs (1936) (abst) Rv Geol et Sci conn 16, fasc 10:578 (1936) (abst) G Zentralbl, Abt A, Bd 58:146 (1936) (abst) Rv Geol et Sci conn 16:578 (1936)

Hoover, J. W.

32 Geographic littoral of northern California. (abst) G Soc Am, B 43:246 (1932)

Hopkins, G. R.

31 (and Backus, H.) Natural gas. U S B M, Min Res, Calendar Year 1929:319-340, California 319, 320, 322, 324, 325, 326, 327, 328, 330, 331, 332, 334, 336, 337, 339, 340 (1932)... 1930:457-481, California 459, 461, 462, 465, 466, 468, 471, 472, 475, 477, 478, 480, 481 (1932)... 1931:349-372, California 351, 352, 358, 360, 363, 366, 369, 372 (1933)... 1932-1933 Min Yb, Stat App:103-116, California 104, 105, 107, 108, 112, 115 (1934)... 31-45, California 32, 33, 36, 39, 42, 44 (1935)

31a (and Seeley, E. M.) Natural gasoline plants in the United States, January 1, 1932. U S B M, Inf Circ 6635:28 pp, California 2, 3, 4, 5, 6, 7, 8 (1932)

31b (and Cochrane, E. W.) Petroleum refineries in the United States January 1, 1931. U S B M, Inf Circ 6485:1-21, California 5, 6, 7-9 (1931)... January 1, 1932. U S B M, Inf Circ 6641:1-21, California 3, 4, 6-9 (1932)... January 1, 1933, U S B M, Inf Circ 6728:1-28 (1933)... January 1, 1934. U S B M, Inf Circ 6807:1-30 (1934)

32 (and Seeley, E. M.) Natural gasoline. U S B M, Min Yb, Stat App 1932-1933:55-65, California 55, 56, 57, 58, 60, 61, 62, 63, 65 (1934)

32a (and Coons, A. B.) Crude petroleum and petroleum products. U S B M, Min Yb, Stat App 1932-1933:299-372, California 299, 300, 302, 303, 304, 305, 306, 309, 311, 327, 328, 330, 331, 333, 334, 335, 336, 338, 339-340, 341, 342, 345, 346, 347, 348, 349, 350, 351, 353, 354, 359, 360, 361,

Hopkins, G. R.—Cont.

362, 363, 364, 365, 368, 369, 371, 372 (1934)

32b Crude petroleum and petroleum products. U S B M, Min Yb 1932-1933: 459-495, California 464, 466, 467, 470, 471-473, 486, 487, 488, 489, 490-493, 494, 495 (1933)... 1934:653-690, California 662, 663, 666, 667, 668-669, 682, 685, 687, 690 (1934)

32c (and **Seeley,** E. M.) Natural gasoline. U S B M, Min Res, Calendar Year 1929:299-318, California 299, 302, 303, 304, 305, 306, 308, 309, 311, 312, 313, 314, 317, 318 (1932)... 1930:433-456, California 435, 436, 437, 438, 439, 440, 443, 444, 445, 446, 447, 449, 450, 451, 453, 455, 456 (1932)... 1931:331-347, California 331, 333, 334, 335, 336, 338, 339, 340, 341, 342, 343, 344, 347 (1933)

32d (and **Coons,** A. B.) Crude petroleum and petroleum products. U S B M Min Res, Calendar Year 1929:421-521, California 421, 423, 424, 427, 428, 430, 431, 432, 434, 435, 437, 439-444, 476, 477, 478, 479, 480, 481, 482, 483, 484, 486, 487, 488, 490, 500-503, 504, 505, 507, 514, 516, 517, 518, 521 (1932)... 1930:775-876, California 775, 777, 778, 779, 781, 782, 784, 785, 786, 788-789, 790, 792-796, 829, 830, 832, 833-834, 835, 836, 837, 838, 840, 841, 856-859, 860, 861, 862, 863, 868, 869, 871, 872, 873 874, 875 (1932)... 1931:553-675, California 553, 554, 559, 560, 562, 564, 566, 567, 568, 570-571, 594, 595, 596, 598, 599, 601, 604, 606, 607, 608, 610, 612, 613, 614, 617, 619, 620, 622, 623, 624, 625, 626, 627, 629, 632, 633, 635, 636, 637, 638, 639, 640, 642, 643, 645, 647, 648, 654, 655, 656, 657, 658, 659, 660, 661, 662, 665, 666, 668, 671, 672, 673, 674, 675 (1933)... Min Yb, (Stat App) 1932-1933:299-372, California 299, 300, 303, 304, 305, 306, 309, 311, 326, 327, 328, 329, 330, 331, 332, 333, 334, 335, 336, 337, 338, 339, 340, 341, 342, 345, 346, 347, 348, 349, 350, 351, 353, 354, 359, 360, 361, 362, 363, 364, 365, 368, 369, 371, 372 (1934)... 1934:199-281, California 199, 200, 201, 203, 205, 206, 207, 208, 209, 210, 213, 215, 230, 231, 232, 333, 234, 235, 236, 237, 238, 240, 242, 244, 245, 246, 247, 248, 250, 252, 253, 254, 255, 256, 257, 258, 259, 260, 262, 263, 268, 269, 270, 271, 272, 273, 274, 276, 277, 278, 279, 280 (1935)

35 Crude petroleum and petroleum products. U S B M, Min Yb 1935:725-758, California 733, 734, 736, 738, 740, 741, 751, 753, 755, 758 (1935)

36 (**White,** A. G.; and **Breakey,** H. A.) Crude petroleum and petroleum products. U S B M, Min Yb 1936:667-723, California 668, 674, 678, 681, 686, 687, 691, 693, 695, 711, 713 (1936)

36a (and **Lott,** F. S.) Natural gas. U S B M, Min Yb 1936:725-748, California 726, 728-729 (1936)

Hopkins, G. R.—Cont.

36b Natural gasoline. U S B M, Min Yb 1936:749-759, California 750, 751, 753, 754, 755 (1936)

Horner, A. C.

33 (and **Wailes,** C. D. Jr.) Earthquake damage analysed by Long Beach officials. Eng News-Rec 110:684-686 (1933) (*abst*) Canada, Dom Obs, Pub, Bib Seism 10:334 (1933) (*abst*) Rv Geol et Sci conn 14:175-176 (1933-1934)

Horner, R. R.

18 Notes on the black sand deposits of southern Oregon and northern California. U S B M, Tech P 196:5-39 (1918)

Horton, F. W.

34 (and **Gaylord,** H. M.) Gold, silver, copper, lead, and zinc in California. U S B M, Min Yb 1934:149-158 (1934)... 1935:147-195 (1935)

Hoskins, E. E.

36 (and **Byerly,** P.) Earthquakes in northern California and the registration of earthquakes at Berkeley, Mount Hamilton, Palo Alto, San Francisco, Ferndale from January 1, 1936, to March 31, 1936. Cal Univ, Seism Sta, B 6:1-37 (1936)... April 1, 1936 to June 30, 1936, B 6:38-85 (1936)

Hough, F. W.

34 Geodetic engineering applied to Colorado river aqueduct. Military Eng 26, no 146:124-129 (1934) (*abst*) Eng Index 1934: 515 (1934)

Howard, Arthur D.

32 Microcrystals of barite from Barstow, California. Am Mineralogist 17: 120-121 (1932) (*ann*) An Bib Ec G 5:33 (1932) Chem Absts 26:4775 (1932)

Howard, Hildegard

31 Pliocene bird remains from Santa Barbara, California. Condor 33:30-31 (1931) (*abst*) Pale Zentralbl 1:68 (1932)

32 Eagles and eaglelike vultures of the Pleistocene of Rancho La Brea (California). Carnegie Inst, Wash, Contr Paleontology, Pub 429:82 pp, 3 figs, 29 pls (1932) (*abst*) Pale Zentralbl 4:114-115 (1934)

32a A new species of cormorant from Pliocene deposits near Santa Barbara, California. Condor 34:118-120, 4 figs (1932)

33 A new species of owl from the Pleistocene of Rancho La Brea, California. Condor 35:66-69, 1 fig (1933) (*abst*) Pale Zentralbl 5:169 (1934)

Howard, P. J.

32 **(Crown,** W. J.; and **Pierce,** G. G.) Developments in the Long Beach oil field. Cal Dp Nat Res, Div Oil and Gas, California Oil Fields 18, no 2:5-26 (1932) (*ann*) An Bib Ec G 7:100 (1935) (*abst*) Rv Geol et Sci conn 14:625 (1934)

Howe, Marshall A.

33 Eocene marine algae (Lithothamnieae) from the Sierra Blanca limestone, Santa Barbara County, California. (*abst*) G Soc Am, Pr 1933:87 (1934) G Soc Am, B 45:507-518 (1934) (*abst*) Rv Geol et Sci conn 15: 522 (1935)

Hubbard, J. D.

17 Drift mining in California. Eng M J 104:863-866 (1917)

Hubbard, Prevost

31 Asphalt. Min Ind 39:56-63, California 57 (1931)... 40:57-62, California 58 (1932)... 41:46-52, California 47 (1933)... 42:50-56, California 51 (1934)... 43:52-58, California 52 (1935)

Huber, Ernst

34 Anatomical notes on Pinnipedia and Cetacea. Carnegie Inst, Wash, Pub 447: 105-136, California 107-112 (1934)

Huber, Walter L.

30 San Francisco earthquakes of 1865 and 1868. Seism Soc Am, B 20:261-272 (1930)

Hughes, H Herbert

31 Iceland spar and optical fluorite. U S B M, Inf Circ 6468:1-18, California 11 (1931)

32 (and **Allan,** M.) Sand and gravel. U S B M, Min Yb, Stat App 1932-1933: 289-297, California 293, 294 (1934)... 1935: 939-948, California 942 (1935)

32a Soapstone. U S B M, Inf Circ 6563: 18 pp (1932) (*abst*) U S B M, List Pub 1910-1932:154 (1933)

32b (and **Bagley,** B. W.) Cement. U S B M, Min Yb 1932-1933: 565-575, California 571 (1933)... 1934:775-797, California 782 (1934)

32c (and **Middleton,** J.) Feldspar. U S B M, Min Res, Calendar Year 1930: 137-149, California 143, 145-146 (1932)... 1931: 179-190, California 184, 186-187 (1933)

35 **(Bagley,** B. W., and **Shuey,** E. T.) Cement. U S B M, Min Yb 1935:883-909, California 887, 898, 901, 904 (1935)

36 (and **Cornthwaite,** M. A.) Sand and gravel. U S B M, Min Yb 1936:841-848. California 844 (1936)

Huguenin, E.

30 Operations in district No. 1, 1929. Cal Dp Nat Res, Div Oil and Gas, California Oil Fields 15, no 3:29-43 (1930)... 1930, 16, no 3:25-34 (1931)... 1931, 17, no 3:17-24 (1932)... 1932, 18, no 3:17-24 (1933)... 1933, 20, no 3:17-24 (1934)

Hulin, Carlton D.

31 A Mother-Lode gold ore. (*abst*) Eng Index 1930:847 (1930) Zs Prak G 39: 48 (1931) Chem Absts 25:1464 (1931)

31a Subsequent faulting in the Great Basin. (*abst*) G Soc Am, B 42:307 (1931) N Jb 1931, Ref 2:446-447 (1931) G Zentralbl 43:407 (1930-1931)

33 Geological relations of ore deposits in California. *In* Ore deposits of the western states. Am I M Eng, Lindgren Volume:240-253 (1933)

35 Geologic features of the dry placers of the northern Mojave desert. Cal Dp Nat Res, Div Mines M G, J, St Mineralogist's Rp 30:417-426 (1935) (*ann*) An Bib Ec G 8:55 (1935) (*abst*) G Zentralbl, Abt A, Bd 57:155 (1936)

Hunziker, A. A.

35 (and **Shrock,** R. R.) A study of some Great Basin lake sediments of California, Nevada, and Oregon. J Sed Petrology 5:9-30 (1935)

Hurlbut, C. S., Jr.

35 Dark inclusions in tonalite from southern California. Am Mineralogist 20: 609-630, 9 figs (1935) (*abst*) 20:205 (1935)

Hutcheson, Robert B.

31 Micropaleontology: an important development in geological science. Oil B 17:119-121 (1931)

Hutchinson, A.

12 On the identity of neocolemanite with colemanite. Miner Mag 16:239-246 (1912)

Huttl, J. B.

33 Short trip to Mother Lode. Eng M J 134:521-522 (1933) (*Abst*) Eng Index 1933: 722 (1933)

34 From the Siskiyous to the South. Mines of the Mother-Lode. Eng M J 135: 506-510 (1934)

34a Rock asphalt. Eng M J 135:116-117, 259 (1934) (*abst*) Eng Index 1934:56 (1934)

35 Natomas' newest dredge. Eng M J 136:270-272 (1935) (*ann*) An Bib Ec G 8:56 (1935)

35a Montezuma-Apex — newest Mother Lode producer. Eng M J 136:173-174 (1935) (*ann*) An Bib Ec G 8:56 (1935) (*abst*) Eng Index 1935:523 (1935)

Huttl, J. B.—Cont.

35b Gold placer mining revived at Lincoln. Eng M J 136:440-443 (1935) (*abst*) G Zentralbl, Abt A, Band 57:415 (1936)

35c Activity spurred in camps of southern California. Eng M J 136:343-347 (1935) (*abst*) G Zentralbl, Abt A, Band 57:416 (1936)

35d Grass Valley gold mines hold their place. Eng M J 136:617-618 (1935) (*ann*) An Bib Ec G 8:290 (1936)

35e Bib Canyon's surface plant an instance of modern efficient design. Eng M J 136:216-219 (1935) (*abst*) Eng Index 1935:523 (1935)

Hyatt, Edward

32 Records of ground-water levels at wells for the year 1934, and precipitation records for the season 1933-1934. Cal Dp Pub Works, Div Water Resources, B 39-C:148 pp (1932) (*ann*) An Bib Ec G 8:162 (1936)

Ickes, E. L.

36 Estimation of probable value of wildcat land. Am As Petroleum G, B 20:1005-1018, California 1006, 1007 (1936)

Irving, John

32 (**Vonsen, M.** and **Gonyer, F. A.**) Pumpellyite from California. Am Mineralogist 17:338-342 (1932) (*abst*) Rv Geol et Sci conn 13:84 (1933) Min Absts 5:233 (1933) Chem Absts 26:4773 (1932)

Jackson, Charles F.

32 (and **Knaebel, J. B.**) Small-scale placer mining methods. U S B M, Inf Circ 6611:17 pp, 25 figs (1932) (*abst*) U S B M, List Pub 1910-1932:156 (1933)

32a (and **Meyer, H. M.**) Mercury. U S B M, Min Res, Min Yb 1932-1933:225-241, California 229, 230, 231-233 (1933)... 1934:385-397, California 390-391 (1934)

32b (and **Knaebel, J. B.**) Gold mining and milling in the United States and Canada. U S B M, B 363:1-143 (1932)

Jaeger, E. C.

33 California deserts. Stanford Univ Press (1933) (*Review*) Geog J 84:267-268 (1934)

Jaggar, T. A.

26 (and **Finch**, R. H.) Tilting and level changes at Pacific volcanoes. Third Pan-Pacific Sc Cong, Tokyo, Pr 1:672-686 (1926) (*abst*) Canada, Dom Obs, Pub, Bib Seism 10:42 (1929)

26a Volcanic relations of great Pacific earthquakes. (*abst*) Third Pan-Pacific Sc Cong Tokyo, Pr 1:370 (1926)

Jaggar, T. A.—Cont.

30 The Lassen myth. The Volcano Letter, no 266:3 January 30 (1930)

31 Hot volcanism of northwest United States and Canada. Volcano Letter, no 334: May 21 (1931)

31a Volcanic waters of Napa County, California. The Volcano Letter, no 340: July 2 (1931)

Jakosky, J. J.

31 Uses of geophysics in gold mining. M J Ariz 15, no 6: 5-7, 37 (1931)

32 (and **Wilson, C. H.**) Use of geophysics in placer mining. M J Ariz 16, no 14:3-4, 29 (1932) (*abst*) Rv Geol et Sci conn 13:416 (1933) (*abst*) Eng Index 1932:617 (1932) G Zentralbl 51:485-486 (1934)

34 (and **Wilson, C. H.**) Geophysical studies in placer and water-supply problems. Am I M Eng, Tech Pub 515:18 pp, pls (1934) Eng M J 135:71-74 (1934) (*abst*) Eng Index 1934:519 (1934)

36 (and **Wilson**, C. H.) Electrical mapping of oil structures. M Metal 17:231-237 (1936)

Jamison, C. E.

10 Santa Clara River placers. M Sc Press 100:360-361 (1910)

Jay, Michael

31 Signal Hill's tenth birthday. Oil B 17:528-529 (1931)

Jeffery, J. A.

31 (with **Woodhouse**, C. D.) Note on a deposit of andalusite in Mono Co., Cal: its occurrence and technical importance. Cal Dp Nat Res, Div Mines, Mining in California, St Mineralogist's Rp 27:459-464 (1931) (*ann*) An Bib Ec G 4:271 (1931)

32 (with **Woodhouse**, C. D.) Mining andalusite in Mono County, California. M J Ariz 15, no 16:5-6, 43-44 (1932)

Jenkins, Olaf P.

30 Geological survey of California under way. (*abst*) Eng Index 1930:822 (1930)

30a Geologic Branch. Progress report. Cal Dp Nat Res, Div Mines, Mining in California, St Mineralogist's Rp 26:330-333, 1 fig (1930)... 26:473-474 (1931)... 27:402-403, 492-493 (1931)... 28:109 (1932)... 29:492-493 (1932)

30b Geologic Branch. Geological and economic mineral survey. Cal Dp Nat Res, Div Mines, Mining in California. St Mineralogist's Rp 26:138-144 (1930)

31 California geological survey. M J Ariz 14, no 21:4 (1931)

31a Stratigraphic significance of the Kreyenhagen shale of California. Cal Dp

Jenkins, Olaf P.—Cont.

Nat Res, Div Mines, Mining in California, St Mineralogist's Rp 27:141-186, 10 figs, 1 pl (1931) (*abst*) G Soc Am, B 42: 363-304 (1931) (*abst*) G Zentralbl 46:195-196 (1932) (*abst*) Eng Index 1931:1036 (1931)

31b Sandstone dikes as conduits for oil migration through shales. (*abst*) Eng Index 1930:1300 (1930) Zs Prak G 39:29 (1931)

31c Geologic Branch. Record of progress by federal departments. Cal Dp Nat Res, Div Mines, Mining in California, St Mineralogist's Rp 27:67-73, 1 fig (1931)

31d Geologic Branch. Publication of papers on the geology of California. Cal Dp Nat Res, Div Mines, Mining in California, St Mineralogist's Rp 27:140 (1931)

32 Compilation of the geology of the Mount Shasta quadrangle, California. (*abst*) Pan Am G 55:361 (1931) (*ann*) An Bib Ec G 5:238 (1932)

32a Report accompanying geologic map of northern Sierra Nevada. Cal Dp Nat Res, Div Mines, Mining in California, St Mineralogist's Rp 28:279-298 (1932) (*ann*) An Bib Ec G 6:9 (1934)

32b Geologic Branch. Contributions to the study of sediments. Cal Dp Nat Res, Div Mines, Mining in California, St Mineralogist's Rp 29:12-13 (1932)

32c Geologic Branch. Current notes. Cal Dp Nat Res, Div Mines, Mining in California, St Mineralogist's Rp 28:278 (1923)... 29:74-75, 340 (1933)... 30:3-4, 117, 328-329, 1 fig (1934)... 31:50, 112, 339, 486 (1935)

33 Use of geology in seeking gold. M J. Ariz 17, no 2:3-4 (1933)

33a The San Francisco peninsula. Int G Cong, Guide Book 16:Exc C-13-18 (1933) (*abst*) N Jb 1936, Ref 3:65-66 (1936)

33b (and others) Middle California and western Nevada. Int G Cong, Guide Book 16, Exc C-1 116 pp. 19 figs (incl geol map) 18 pls (1933) (*abst*) G Zentralbl Abt A, Bd 53:240 (1934)

34 Resurrection of early surfaces in the Sierra Nevada. Cal Dp Nat Res, Div Mines, M G, J, St Mineralogist's Rp 30: 5-6 (1934) (*abst*) G Soc Am, Pr 1934:338 (1935) (*ann*) An Bib Ec G 7:243 (1934) (*abst*) G Zentralbl, Abt A, Bd 55:457 (1935)

34a (and **Wright,** W. Q.) California's gold-bearing Tertiary channels. Eng M J 135:497-502 (1934) (*abst*) G Zentralbl, Abt A, Bd 55:457 (1935)

35 New technique applicable to the study of placers. Cal Dp Nat Res, Div Mines, M G, J, St Mineralogist's Rp 31: 143-211 (1935)

Jenkins, Olaf P.—Cont.

35a Progress of the geologic map of California. Am As Petroleum G, B 19: 135 (1935) (*abst*) Pan Am G 63:314-315 (1935) (*abst*) G Soc Am, Pr 1935:337 (1936)

35b Preliminary legend of rock formation units used on the new geologic map of California. Cal Dp Nat Res, Div Mines, M G, J St Mineralogist's Rp 31: 113-114 (1935)

Jenny, W. P.

32 Magnetic vector study of regional and local geologic structure in principal oil states. Am As Petroleum G, B 16: 1177-1203, 10 figs (1932)

34 Structural trends on the gulf coast. Oil Weekly 74:33-40, California 40, figs (1934)

Jensen, Joseph

30 (**McDowell,** G.; **Goold,** W. D.; and **Gwin,** M. L.) Deep sand development at Santa Fe Springs. *In* Petroleum Dev and Tech, Am I M Eng, Tr 1930:310-321 (1930)

30a (and **Stevens,** J. B.) Water invasion, McKittrick oil field. M Metal 11: 470-471 (1930)

30b Unit operation in California. *In* Petroleum Dev and Tech, Am I M Eng, Tr 1930:69-79 (1930)

31 Unit operation in California, with discussion of Kettleman North Dome Association. *In* Petroleum Dev and Tech, Am I M Eng, Tr 1931:80-91 (1931)

31a (and **Hertel,** F. W.) Development in a part of the Ventura Avenue oil field. *In* Petroleum Dev and Tech, Am I M Eng, Tr 1931:149-156 (1931) Oil Weekly 59, no 6:34-43 (1930) M Metal 11: 475-478 (1930)

31b (and **Stevens,** J. B.) Water problems of McKittrick oil field. *In* Petroleum Dev and Tech, Am I M Eng, Tr 92:164-167 (1931)

32 Oil curtailment in California. M Met 13:521-522 (1932)

33 Kettleman Hills Middle Dome unit plan. *In* Petroleum Dev and Tech. Am I M Eng, Tr 1933:160-167 (1933)

34 California oil field waters. *In* Problems of petroleum geology, Sidney Powers Memorial Volume. Am As Petroleum G, Pub:953-985 (1934)

Johnson, A. M.

32 (and **Hodges,** F. C.) Subsurface storage of oil and gas in the Brea-Olinda and Lompoc fields. Cal Dp Nat Res, Div Oil and Gas, Cal Oil Fields 17, no 4:5-12 (1932) (*abst*) Rv Geol et Sci conn 14:73-74 (1934)

Johnson, Bertrand L.

36 (and **Cornthwaite**, M. A.) Barite and barium products. U S B M, Min Yb 1936:997-1006. California 998, 999 (1936)

36a (and **Davis**, A. E.) Abrasive materials. U S B M, Min Yb 1936:877-894, California 878, 880, 883 (1936)

Johnson, C. H.

34 (and **Gardner**, E. D.) Placer mining in the western United States. U S B M, Inf Circ 6786:1-74... 6787:1-89... 6788:1-81 (1934)

Johnson, Floyd L.

34 (and **Barbat**, W. F.) Stratigraphy and foraminifera of the Reef Ridge shale, Upper Miocene, California. J Paleontology 8, no 1:3-17, 1 pl (1934) (*abst*) Pan Am G 59:239 (1933) N Jb 1934, Ref 3:624 (1934) Biol Absts 9:1396, entry 12523 (1935)

Johnson, Fred W.

35 (and **Gardner**, E. D.) Prospecting for lode gold and locating claims of the public domain. U S B M, Inf Circ 6843: 1-18, California 2, 6, 11, 14, 16-18 (1935)

Johnston, J. H.

35 Geologist contends Nevada Lakes caused by gigantic gas blow out: oil possibilities at Nevada Lakes. Cal Oil World 27, no 24:8-9 (1935)

Johnston, William D., Jr.

32 Geothermal gradient at Grass Valley, California. Wash Ac Sc, J 22:267-271, 1 fig (1932) (*abst*) G Zentralbl Abt A, 48:22 (1932)

32a Geothermal gradient of the Mother Lode belt, California: a reply. Wash Ac Sc, J 22:390-393 (1932) (*abst*) G Zentralbl, Abt A 48:228 (1932)

32b Structure of the Grass Valley batholith, California. (*abst*) Wash Ac Sc, J 22:317-318 (1932)

33 Grass Valley district. *In* Ore deposits of the western states. Am I M Eng, Lindgren Volume: 580 (1933)

33a Copper in Trinity County, California. XIV Int G Cong, Wash, Copper resources of the world, 1933:251 (1933)

34 (and **Cloos**, E.) Structural history of the fracture systems of Grass Valley, California. Ec G 29:39-63 (1934) (*ann*) An Bib Ec G 7:41 (1935) (*abst*) Rv Geol et Sci conn 14:180 (1934) G Soc Am, B 44:88 (1933) (*abst*) G Zentralbl Abt A 48:186-188 (1932) (*abst*) N Jb 1934, Ref 2:523-524 (1934)... 1935, Ref 2:156 (1935) (*abst*) Eng Index 1934:533 (1934) (*abst*) G Zentralbl, Abt A, Bd 53:348-349 (1936)

Johnston, William D., Jr.—Cont.

36 Nodular, orbicular, and banded chromite in northern California. Ec G 31:417-427 (1936)

Joleaud, M. L.

31 Les recents progrés de nos connaissances sur l'histoire du Pacifique aux temps tértiatres et la théorie de Wegener. Ac Sc Paris, C R 192:628-629 (1931)

Jomercq, J. Junr

35 (**Mejea**, J.; and **Uren**, L. C.) Large diameter wells are indicated. Petroleum World (London) 32:260-262 (1935)

Jones, Austin E.

27 (and **Byerly**, P.) Registration of earthquakes at the Berkeley Station and at the Lick Observatory Station from April 1, 1926 to March 31, 1927. Cal Univ, Seism Sta, B 2:202-250 (1927)

Jones, P. H.

35 (**Hoots**, H. W.; and **Blount**, A. L.) Marine oil shale, source of oil in Playa del Rey field, California. Am As Petroleum G, B 19:172-206 (1935) (*ann*) An Bib Ec G 8:142 (1935)

Jones, Paul L.

36 Testing placer ground in a unique way. Eng M J 137:337-338 (1936)

Jordan, D. S.

31 New sharks from the Temblor group in Kern County, California. (*abst*) N Jb 1931, Ref 3:293 (1931)

Josephson, W. G.

32 Argonaut mine of today. M Metal 13:475-477 (1932) (*abst*) Eng Index 1932: 634 (1932)

Judson, S. A.

30 (and **Murphy**, P. C.) Deep sand development in Barbers Hill. Am As Petroleum G, B 14:719-741, 11 figs (1930) Oil Gas J 28, no 44:146-147 (1930) Oil Weekly 57 (3):25-30, 76, 78 (1930) (*abst*) Eng Index 1930:1300 (1930)

Julihn, C. E.

32 (with **Meyer**, H. M.) Copper. U S B M, Min Res, Calendar Year 1929:525-580, California 527, 538, 539, 540, 541, 542, 543, 544, 545, 557-558 (1932)... 1930:691-748, California 725 (1933)... 1931:575-600, California 582, 583, 584, 585, 586, 587, 590, (1934)... Min Yb 1932-1933:27-52, California 30, 32, 33, 34, 35, 36, 37 (1933)... 1934:53-77, California 63-64 (1934)

Jullum, Henry.

32 Milling at the Argonaut. M Metal 13:476-477 (1932)

Jumper, Harry D.

33 Mining, treatment methods and cost at the plant of the Consolidated Rock Products Co., Durbin, California. U S B M, Inf Circ 6607:1-20 (1933)

Justice, C. W.

33 (and Bowles, O.) Growth and development of the non-metallic mineral industries. U S B M, Inf Circ 6687:1-50, California 10-12, 29-36, 40-42, 44-46, 49 (1933)

Kaplow, E. J.

33 (and Dodd, H. V.) Kettleman North Dome and Kettleman Middle Dome fields —progress in development. Cal Dp Nat Res, Div Oil and Gas, California Oil Fields 18, no 4:5-20 (1933) (ann) An Bib Ec G 7:324 (1934)

Keen, A. Myra

35 (and Schenck, H. G.) West American marine molluscan provinces. (abst) Pan Am G 63:375-376 (1935) (abst) G Soc Am, Pr 1935:413 (1936)

Keenan, Marvin Francis

32 The Eocene Sierra Blanca limestone at the type locality in Santa Barbara County, California. San Diego Soc N H, Tr 7:53-84, 4 figs, 3 pls (1932) (abst) G Zentralbl, Abt A 48:395 (1932) Pale Zentralbl 3:103 (1933)

Keith, Arthur

28 Structural symmetry in North America. G Soc Am, B 39:321-385 (1928) (abst) N Jb 1931, Ref 3:555-557 (1931)

Kelley, V. C.

34 Westernmost Santa Monica Mountains. (abst) Pan Am G 61:308-309 (1934) (abst) G Soc Am, Pr 1934:311-312 (1935)

35 (and Soske, J. L.) Wave-built pumice deposits and Salton rhyolitic hills. (abst) Pan Am G 63:319-320 (1935) (abst) G Soc Am, Pr 1935:341 (1936)

36 (and Soske, J. L.) Origin of the Salton volcanic domes, Salton Sea, California, J G 44:496-509, figs (1936) (abst) Rv Geol et Sci conn 16:514-515 (1936) (abst) G Zentralbl, Abt A, Bd 58:28 (1936) (abst) Rv Geol et Sci conn 16:514-515 (1936)

36a Occurrence of claudetite in Imperial County, California. Am Mineralogist 21:137-138 (1936)

Kelley, W. P.

33 (Woodford, A. O.; and Brown, S. M.) Clay minerals of California soils. (abst) Pan Am G 59:315-316 (1933)

Kellogg, A. E.

30 Hydraulic concentration of cinnabar ore. M J Ariz 14, no 15:13, 36, 3 figs (1930) (abst) Eng Index 1931:865 (1931)

31 Origin of flour gold in black sands. M J Ariz 14, no 20:3-4, 49-50 (1931)

Kellogg, Remington

31 Pelagic mammals from the Temblor formation of the Kern River region, California. Cal Ac Sc, Pr (4) 19:217-397, 134 figs (1931) (abst) G Zentralbl 44:454-455 (1931) Pale Zentralbl 2:324 (1933)

32 A Miocene long-beaked porpoise from California. Smiths Misc Col 87, no 2:11 pp, 4 pls (1932) (abst) Pale Zentralbl 4: 234 (1934) Biol Absts 7:1257, entry 12523 (1933)

34 A new cetothere from the Modelo formation at Los Angeles, California. Carnegie Inst, Wash, Pub 447:83-104, 3 figs, 1 pl (1934) (abst) N Jb 1934, Ref 3:471 (1934)

34a Study of the skull of a fossil sperm whale from the Temblor Miocene of southern California. Biol Absts 8:282, entry 2550 (1934)

34b Fossil pinnipeds from California. Biol Absts 8:282, entry 2551 (1934)

Kelly, Junea W.

33 Limestone weathering and plant associations of the San Francisco region. Cal Dp Nat Res, Div Mines, M G, J, St Mineralogist's Rp 29:362-367 (1934)

Kemnitzer, W. J.

31 (and Arnold, R.) Petroleum in the United States and Possessions. 1052 pp, maps, charts, tables (1931)

Kendall, E. B.

33 Mining methods and costs at Plant C, Eliot, California, of the Pacific Coast Aggregates, Inc. U S B M, Inf Circ 6705: 12 pp, 11 figs. (1933) (abst) U S B M List Pub July 1, 1932 to June 30, 1933, Suppl:16 (1933)

Kennard, T. G.

33 (and Rambo, A. I.) Occurrence of rubidium, gallium, and thallium in lepidolite from Pala, California. Am Mineralogist 18:454-455 (1933) (ann) An Bib Ec G 6:194 (1934)

35 Spectrographic examination of smoky and ordinary quartz from Rincon, California. Am Mineralogist 20:392-399 (1935) (abst) N Jb 1935, Ref 1:445 (1935) (abst) Rv Geol et Sci conn 15:295 (1935)

Kennedy, H. S.

35 (and **Fowler**, H. C.) Natural gas. U S B M, Min Yb 1935: 795-819, California 798-801 (1935)

Kerr, Paul Francis

31 Bentonite from Ventura, California. Ec G 26:153-168 (1931) (*ann*) An Bib Ec G 4:68 (1931) (*abst*) G Zentralbl 45:475 (1931) Zs Prak G 39:194 (1932) Chem Absts 25:3275 (1931)

31a (**Schenck**, H. G.; and **Muller**, S.) Geology of the Ventura quadrangle, California. G Soc Am, B 42:186-187 (1931) (*abst*) Pan Am G 55:64 (1931)

31b (and **Schenck**, H. G.) Significance of the Matilija overturn. (*abst*) N Jb 1931, Ref 3:544-545 (1931)

32 Occurrence of andalusite and related minerals at White Mountain, California. Ec G 27:614-643, 18 figs (1932) (*ann*) An Bib Ec G 5:299 (1932) (*abst*) N Jb 1933, Ref 2:289 (1933) Zs Prak G 41:167 (1933) Min Absts 5:288 (1933) Chem Absts 27:2114 (1933) (*abst*) Rv Geol et Sci conn 15:51-52 (1935)

32a (and **Ross**, C. S.) Manganese minerals of a vein near Bald Knob, North Carolina. Am Mineralogist 17:1-18, 2 figs, California 13 (1932) Min Absts 5:50-51 (1932)

36 (and **Cameron**, E. N.) Fuller's earth of bentonitic origin from Tehachapi, California. Am Mineralogist 21, no 4:230-237 (1936) (*abst*) Rv Geol et Sci conn 16:435 (1936)

Kew, W. S. W.

27. Geologic formations of a part of southern California and their correlation. (*abst*) N Jb 1927, Ref 2, Abt B:114 (1927)

32 Los Angeles to Santa Barbara. Int G Cong, Guide Book 15:48-68 (1932) (*abst*) N Jb 1936, Ref 3:61-62 (1936)

32a (and **Brown**, A. B.) Occurrence of oil in metamorphic rocks of San Gabriel Mountains, Los Angeles County, California. Am As Petroleum G, B 16:237 (1933) G Zentralbl Abt A 48:317 (1932) Inst Pet Tech, J 18:415A (1932)

32b (and **Woodring**, W. P.) Tertiary and Pleistocene deposits of the San Pedro Hills, California. (*abst*) Wash Ac Sc, J 22, no 2:39-40 (1932)

Keyes, Charles Rollin

31 Barrelian series in eastern California. Pan Am G 56:76-78 (1931)

31a Mojave as geological title. Pan Am G 56:155-158 (1931)

31b Esmeralda formation of California. Pan Am G 56:158-160 (1931)

31c Greenwater volcanics around Death Valley. Pan Am G 56:315-316 (1931)

Keyes, Charles Rollin—Cont.

31d Proper usage of the terranal title Camulos in California. Pan Am G 56: 74-76 (1931)

31e Tills of the last glacial cycle. Pan Am G 55:367 (1931)

32 Last glacial cycle and its tills. (*abst*) G Soc Am, B 43:231 (1932)

32a Mechanics of geographic overthrusting. (*abst*) G Soc Am,, B 43:231-232 (1932)

Khomenko, I. P.

30 On the age of the Tertiary formation along the coast of Korf Gulf Kamchatka. Far-east Geological and Prospecting Trust of U. S. S. R., Tr 1933, Fascicle 287:1-32, California 31, 32 (1933) (*abst*) Eng Index 1935:784 (1935)

Khomenko, J. P.

31 Stratigraphy of the Tertiary beds of the Eastern Sakhalin oilfield. 1. The Nutovo and Supranutovo series. Geological and Prospecting Service of the U. S. S. R., Tr 1931, Fascicle 79:1-126, California 113 (1931)

Kiessling, O. E.

31 (and **Herlihy**, K. V.) Clay. U S B M, Min Res, Calendar Year 1931:237-242, California 239 (1933)

33 Minerals yearbook 1932-1933. U S B M, Dp Commerce xiii plus 819 pp, 90 figs (1933) Min Absts 6:26 (1935)

34 (and **Davis**, H. W.) Iron ore, pig iron, ferro-alloys and steel. U S B M, Min Yb 1934:317-366, California 351 (1934)

35 Minerals yearbook 1935. (Compiled under supervision of O. E. Kiessling) Introduction by Kiessling. xiv 1293 pp (1935)

36 (**Davis**, H. W., and **Herring**, C. T.) Iron ore, pig iron, ferro-alloys and steel. U S B M, Min Yb 1936: 365-394, California 375 (1936)

Kimberlin, C. L.

33 (and **Clark**, C. L.) Bottom hole pressure work at Kettleman. Petroleum World 1933 (April):19-20 (1933)

King, Philip B.

32 An outline of the structural geology of the United States. Mt G Cong, Guide Book 28:57 pp, map (1932)

Kirby, J. M.

34 (and **Crook**, T. H.) The Capay formation. (*abst*) Pan Am G 61:377 (1934) (*abst*) G Soc Am, Pr 1934:334 (1935)

35 Geology of the Vacaville-Rumsey Hills area, Solano, Yolo, and Colusa Counties, California. (*abst*) Am As Petroleum G, B 19:1841 (1935)

Kirkham, Virgil R. D.

31 Review of Allen's "The Ione formation of California." J G 39:693-694 (1931)

Kissock, Alan

34 Molybdenum. Min Ind 42:404-409, California 408 (1934)... 43:417-420, California 420 (1935)

Kleinpell, Robert Missen

30 Zonal distribution of the Miocene foraminifera in Reliz Canyon, California. Stanford Univ, Micropaleontology B 2:33-38 (1930)

32 Miocene foraminifera from Reliz Canyon (Monterey County, California). (abst) Pan Am G 58:77 (1932) (abst) G Soc Am, B 44:165 (1933) (abst) Pale Zentralbl 4:149 (1934)

33 Miocene foraminifera from Contra Costa County. (abst) Pan Am G 59:375-376 (1933) (abst) G Soc Am, Pr 1933 1: 390 (1934)

34 Proposed biostratigraphic classification of California Miocene. (abst) Pan Am G 62:76-77 (1934) (abst) G Soc Am, Pr 1934:390-391 (1935)

34a (and **Cushman,** J. A.) New and unrecorded foraminifera from the California Miocene. Cushman Lab Foram Res, Contr 10:1-23 (1934) (abst) N Jb 1934, Ref 3:623-624 (1934) (abst) Rv Geol et Sci conn 14:504 (1934)

34b Miocene foraminifera from Reliz Canyon, Monterey County, California. Stanford University, Abstract Dissertations 9:100-101 (1933-1934)

34c Difficulty of using cartographic terminology in historical geology. Am As Petroleum G, B 18:374 (1934)

35 (**Bramlett,** M. N.; and **Woodring,** W. P.) Stratigraphy and paleontology of the Palos Verdes Hills. Am As Petroleum G, B 19:1842 (1935) (abst) G Zentralbl, Abt A, Bd 57:51-52 (1936) (abst) Rv Geol et Sci conn 16:378 (1936)

35a (and **Schenck,** H. G.) Foraminifera from Gaviota formation. (abst) Pan Am G 64:76 (1935)

36 (and **Schenck,** H. G.) Refugian stage of Pacific Coast Tertiary. Am As Petroleum G, B 20:215-225 (1936)

Kleinpell, W. D.

34 (and **Cunningham,** G. M.) Importance of unconformities to oil production in the San Joaquin Valley, California. In Problems of petroleum geology. Am As Petroleum G, Sidney Powers Memorial Volume:785-805 (1934) (ann) An Bib Ec G 7:314 (1934)

Knaebel, John B.

31 The veins and crossings of the Grass Valley district, California. Ec G 26:375-

Knaebel, John B.—Cont.

398 (1931) (ann) An Bib Ec G 4:49 (1931) (abst) Rv Geol et Sci conn 12:478 (1932) G Zentralbl 46:12 (1932) (abst) Eng Index 1931:672 (1931) (abst) N Jb 1932, Band 2, Ref 2:38 (1932)

32 (and **Jackson,** C. F.) Small-scale placer-mining methods. U S B M, Inf Circ 6611:17 pp, 25 figs (1932) (abst) U S B M, List Pub 1910-1932:156 (1933)

32a (and **Jackson,** C. F.) Gold mining and milling in the United States and Canada. U S B M, B 363:1-143 (1932)

Knapp, Arthur

31 Petroleum and petroleum products. Min Ind 39:431-456, California 433, 434, 435, 444 (1931)... 40:392-415, California 394-396, 404 (1932)... 41:376-394, California 385, (1933)... 42:418-440, California 426 (1934)... 43:429-447, California 434 (1935)

Knopf, Adolph

27 A geologic reconnaissance of the Inyo Range and the eastern slope of the southern Sierra Nevada, California, with a section on the stratigraphy of the Inyo Range, by Edwin Kirk. (abst) N Jb 1927, Ref 1, Abt A:212-213 (1927)

31 The Mother-Lode system of California. (abst) N Jb 1931, Ref 2:445-446 (1931) M Metal 11:23 (1930)

31a (and **Anderson,** C. A.) The Engels copper deposits, California. (abst) Eng Index 1930:450 (1931) (abst) N Jb 1931, Ref 2:439-440 (1931) (abst) Zs Prak G 39:30 (1932) Chem Absts 25:265 (1931)

32 Geothermal gradient of the Mother Lode belt, California. Wash Ac Sc, J 22: 389-390 (1932)

33 The Mother Lode system. Int G Cong Guide Book 16:45-60 (1933) (abst) N Jb 1934, Ref 2:231 (1934)

33a Pyrometasomatic deposits. In Ore deposits of the western states. Am I M Eng, Lindgren Volume:537-557, California 538, 542-543, 552-553, 555, 557 (1933) (abst) N Jb 1934 Ref 2:679 (1934)

33b The Darwin silver-lead mining district, California. (abst) In Ore deposits of the western states, Lindgren Volume. Am I M Eng:552 (1933)

33c Tungsten deposits of northwestern Inyo County, California. (abst) In Ore deposits of the western states. Lindgren Volume Am I M Eng:555 (1933)

35 The Plumas County copper belt. XVI Int G Cong 1:241-245 (1935)

Kock, Thomas W.

33 Analysis and effects of current movement on an active fault in Buena Vista Hills oil field, Kern County, California. Am As Petroleum G, B 17:694-712 (1933) (ann) An Bib Ec G 6:97

Kock, Thomas W.—Cont.

(1934) (*abst*) Inst Pet Tech, J 19:395A
(1933) (*abst*) Rv Geol et Sci conn 14:106
(1934)

Kraemer, A. J.

31 Properties of California crude oils.
U S B M, Rp Invest 3074:12 pp (1931)
(*abst*) U S B M, List Pub 1910-1932:115
(1933)

Kraut, Max

32 Floating gold on the Mother Lode.
M Metal 13:175-176 (1932)

Kudlich, R. H.

32 (and **Herlihy,** K. V.) Clay. U S
B M, Min Yb 1932-1933:639-645, California 641 (1933)

Kunz, George F.

03 (and **Baskerville,** C.) Kunzite and
its unique properties. Am J Sc (4) 18:25-
28 (1904)
30 Platinum group metals. Min Ind
39:474-491, California 476 (1931)... 40:430-
444, California 430 (1932)... 41:430-444,
California 430 (1933)

Lahee, Frederic H.

34 A study of the evidences for lateral
and vertical migration of oil. *In* Problems of petroleum geology. Am As Petroleum G, Sidney Powers Memorial Vol-
ume:399-431 (1934)
34a (and **Washburne,** C. W.) Oil field
waters (foreword) *In* Problems of petroleum geology. Am As Petroleum G, Sidney Powers Memorial Volume:833-840
(1934)
34b (and **Wrather,** W. E. Ed) Problems of petroleum geology: a symposium.
Am As Petroleum G, Sidney Powers Memorial Volume (1934) (*Review*) Ec G B
30:194-196 (1935) Inst Petroleum Tech,
J 20:1113-1115 (1934)

Laiming, Boris

31 (and **Cushman,** J. A.) Miocene foraminifera from Los Sauces Creek, Ventura County, California. J Paleontology
5, no 2:79-120, 6 pls, 5 figs (1931) Biol
Absts 9:457, entry 3895 (1935) (*abst*) Pale
Zentralbl 3:187 (1933)

Laizure, C. McK.

32 Elementary placer mining methods
and gold-saving devices. Cal Dp Nat
Res, Div Mines, Mining in California, St
Mineralogist's Rp 28:112-204 (1932)
33 Booming. Cal Dp Nat Res, Div
Mines, M G, J, St Mineralogist's Rp 29:
368-371 (1933) (*abst*) Eng Index 1934:540
(1934)

Laizure, C. McK.—Cont.

34 Elementary placer mining in California and notes on the milling of gold
ores. Cal Dp Nat Res, Div Mines, M G,
J, St Mineralogist's Rp 30:121-281 (1934)
(*abst*) Eng Index 1934:540 (1934)
35 Current mining activities in the San
Francisco district with special reference
to gold. Cal Dp Nat Res, Div Mines, M
G, J, St Mineralogist's Rp 31:24-49 maps
(1935)

Lalicker, Cecil G.

35 New Tertiary Textulariidae. Cushman Lab Foram Res, Contr 11:29-51, California 40, 41, 49 (1935)
35a New Cretaceous Textulariidae.
Cushman Lab Foram Res, Contr 11:1-13,
California 4, 5, 7, 8, pls (1935)

Lancaster, H. K.

32 (and **McKenzie,** M. R.) Milling
methods at the Concentrator of the
Walker Mining Co., Walkermine, California. U S B M, Inf Circ 6555:1-11 (1932)

Lang, W. B.

34 (and **Mansfield,** G. R.) The Texas-
New Mexico potash deposits. Univ Texas, B 2, 3401:641-832, California 642, 643,
652, 653, 676 (1934)

Langley, R. W.

10 (and **Foote,** H. W.) On an indirect
method for determining columbium and
tantalum. Am J Sc (4) 30:393-400 (1910)

Larsen, Esper S.

33 (and **Dunham,** K. C.) Tilleyite, a
new mineral from the contact zone at
Crestmore, California. Am Mineralogist
18:469-473 (1933) Min Absts 5:387 (1934)
(*abst*) N Jb 1934, Ref 1:142 (1934) Chem
Absts 28:1957-1958 (1934)
35 (and **Miller,** F. S.) The Rosiwal
method and the modal determination of
rocks. Am Mineralogist 20:260-273, 7
tables (1935)

Larson, L. M.

30 Osteology of the California road-
runner, Recent and Pleistocene. Cal
Univ Pub Zool 32:409-428, 3 figs (1930)
(*abst*) Pale Zentralbl 2:265 (1933) (*abst*)
N Jb 1933, Ref 3:207 (1933)

Laudermilk, J. D.

30 (and **Woodford,** A. O.) Soda-rich
anthophyllite asbestos from Trinity
County, California. Am Mineralogist 15:
259-262 (1930) Min Absts 5:45 (1932)
Chem Absts 25:1764 (1931)
31 A mineralogical occurrence of iron
tannate. Rocks and Minerals 6:24-25
(1931)

Laudermilk, J. D.—Cont.

31a On the origin of desert varnish. Am J Sc (5) 21:51-66, 14 figs (1931) Min Absts 4:515 (1931) (*Review*) Sc Progress 27:582-583 (1933)

32 (and **Woodford**, A. O.) Concerning Rillensteine. Am J Sc (5) 23:135-154, figs (1932)

32a (and **Woodford**, A. O.) Rilled limestone. (*abst*) G Soc Am, B 43:227 (1932)

33 (and **Woodford**, A. O.) California occurrence of montmorillonite after feldspar. (*abst*) Pan Am G 59:315 (1933) (*abst*) G Soc Am, Pr 1933 1:313 (1934) (*abst*) Rv Geol et Sci conn 14:375 (1933-1934)

34 (and **Woodford**, A. O.) Secondary montmorillonite in a California pegmatite. Am Min 19:260-267 (1934) (*abst*) N Jb, 1934, Ref 1:464 (1934) Chem Absts 28:6089 (1934) (*ann*) An Bib Ec G 7:218 (1934)

35 (and **Woodford**, A. O.) Black iron sulphide in California crystalline limestone. (*abst*) Pan Am G 63:320 (1935) (*abst*) G Soc Am 1935:342 (1936)

36 (and **Merriam**, R.) Two diopsides from southern California. Am Mineralogist 21:715-718 (1936)

Lawrie, H. N.

30 Gold and silver. Min Ind 39:229-294, California 252-253, 258-261 (1931)... 40: 185-257, California 223-225, (1932)... 41: 175-260, California 218-219 (1933)... 42:190-292, California 239-240, 246-247 (1934)... 43:186-294, California 232-233, 237-239 (1935)

36 Gold. Eng M J 137:54-55, 61-62 (1936)

Lawson, Andrew C.

31 The Cordilleran shield. Third Pan Pacific Sci Cong, Tokyo, Pr:371-388 (1928) (*abst*) N Jb 1931, Ref 2:305 (1931)

32 Rain wash erosion in humid regions. G Soc Am, B 43:703-724 (1932)

32a Insular arcs, foredeeps, and geosynclinal seas of the Asiatic Coast. G Soc Am, B 43:353-381, California 363, figs (1932)

33 The geology of middle California. Int G Cong, Guide Book 16:1-12 (1933) (*abst*) N Jb 1936, Ref 3:62-65 (1936)

36 The Sierra Nevada in the light of isostasy. G Soc Am, B 47:1691-1712 (1936)

Leach, C. E.

32 (with **Menken**, F. A.) Overturned plunge on overturned folds in Sespe-Piru Creek district, California. Am As Petroleum G, B 16:209-212 (1932) (*ann*) An Bib Ec G 5:140 (1932) (*abst*) Eng Index 1932:613 (1932)

Leaver, E. S.

31 (and **Woolf**, J. A.) Re-treatment of Mother Lode, California carbonaceous slime tailings. U S B M, Tech P 481:20 pp (1931)

Lee, Bourke

30 Death Valley. X, 211 pp (1930)

Leith, Kenneth

35 (and **Liddell**, Donald M.) The mineral reserves of the United States and its capacity for production. Prepared for the Planning Committee for Mineral Policy. Nat Resources Committee, Wash, D. C. 246 pp, 25 pls (March, 1935)

Lemmon, Dwight W.

35 Augelite from Mono County, California. Am Mineralogist 20:664-668 (1935) (*ann*) An Bib Ec G 8:266 (1936)

Les Journaux Francais

30 Tremblement de terre en Californie. Les journaux francais du 5 Mars 1930. (*abst*) Rv Geol et Sci conn 12:188 (1932)

Leverett, F.

29 Pleistocene glaciations of the northern hemisphere. Science n s 69:231-239, California 236 (1929)

Levorsen, A. Irving

33 Studies in paleogeology. Am As Petroleum G, B 17:1107-1132, California 1113, 1126-1127, 1129 (1933)

Lewis, W. Scott

33 Occurrences of opal in California. Rocks & Min 8:36 (1933)

Leypoldt, Harry

34 Earth-movements in California determined from apparent variation in tidal datum planes. Seism Soc Am, B 24:63-68 (1934)

Liddell, Donald M.

35 (with **Leith**, Kenneth) The mineral reserves of the United States and its capacity for production. Prepared for the Planning Committee for Mineral Policy. Nat Resources Committee, Wash, D. C. 246 pp. 25 pls (March, 1935)

Lilley, Ernest R.

36 Economic geology of mineral deposits. 811 pp, figs, tables. Barite 727, 729. Bentonite 718. Boron 62, 721, 722. Coal 247. Chromite 23, 425. Copper 473. Diatomite 709. Feldspar 701. Fuller's earth 717. Gold 51, 52, 53, 65, 102, 568, 570, 573, 575. Granite 152. Graphite 712. Gypsum 189. Kyanite and andalu-

Lilley, Ernest R.—Cont.

site 770. Magnesite 749, 750. Marble 155. Mercury 59, 550, 553, 554. Natural gas 354. Natural sodium compounds 774. Nitrate 681. Petroleum 296, 297, 298, 308, 309, 326. Potash 668, 676. Pumice 772. Pyrite 644. Salt 660. Silver 600. Slate 156. Strontium 774. Talc 756. Tin 522. Trap rock 512. Tungsten 431. Volcanic rock 138. (1936)

Lindgren, Waldemar

33 Differentiation and ore deposition, Cordilleran region of the United States. *In* Ore deposits of the western states. Am I M Eng, Lindgren Volume:152-180 (1933)

Linton, Robert

36 Clay mining in California. M Metal 17:198-200 (1936)

Livingston, Alfred, Jr.

33 (and **Putnam,** W. C.) Geological journeys in southern California. Los Angeles Jr. Coll, Pub 1:104 pp, 56 figs, 1 pl (1933)

Locke, Augustus

31 Ore hunting in California. M Metal 12:540-541 (1931)
33 (and **Billingsley,** P.) Tectonic position of ore districts in the Rocky Mountain region. Am I M Eng, Tech Pub 501: 12 pp. (1933) (*ann*) An Bib Ec G 2:199-200 (1934)
34 (**Billingsley,** P.; and **Schmitt,** H. A.) Some ideas on the occurrence of ore in the western United States. Ec G 29: 560-576 (1934)

Loel, Wayne

27 (and **Arnold,** R.) New oil fields of the Los Angeles basin, California. (*abst*) N Jb, Abt B, Ref 2:113 (1927)
31 (and **Corey,** W. H.) Geologic history of the Vaqueros period in California. Petroleum World 28, no 8:55, 77, 1 fig (1931)
32 (and **Corey,** W. H.) The Vaqueros formation, lower Miocene of California. I Paleontology. Cal Univ Dp G Sc, B 22:31-410, 61 pls, 2 maps (1932) (*abst*) Rv Geol et Sci conn 13:523-524 (1933) (*abst*) Am J Sc 26:528-529 (1933) (*abst*) Eng Index 1933:536 (1933) (*abst*) N Jb 1935, Ref 3:283-287 (1935)

Logan, Clarence A.

24 Notes on mining during the year 1923. Cal St Mineralogist's Rp 20:1-23 (1924)
30 Sacramento field division. Cal Dp Nat Res, Div Mines, Mining in California,

Logan, Clarence A.—Cont.

St Mineralogist's Rp (Nevada Co) 26:90-137 (1930)... (Yuba County) 26:186-201 (1930)... 27:246-261 (1931) (Butte County) 26:360-412 (1931)... (Alpine County) 27: 488-491 (1931) (*abst*) Eng Index 1930:1114 (1930)... 1931: 898 (1931)
35 Mother Lode gold belt of California. Cal Dp Nat Res, Div Mines, B 108:219 pp, maps (1935) (*ann*) An Bib Ec G 8:56 (1935) (*abst*) G Zentralbl, Abt A, Bd 57:156-157 (1936)
35a Review of gold mining in east-central California. Cal Dp Nat Res, Div Mines, M G, J, St Mineralogist's Rp 31:1-23 (1935)
36 Gold mines of Placer County. Cal Dp Nat Res, Div Mines, M G, J, St Mineralogist's Rp 32:7-96 (1936)
36a (and **Franke,** H. A.) Mines and mineral resources of Calaveras County. Cal Dp Nat Res, Div Mines, M G, J, St Mineralogist's Rp 32:226-364 (1936)

Logan, J.

28 Geophysics reveal vast petroleum deposits in coast region. Oil Weekly 51, no 9:40-50 (1928)

Lohman, Kenneth E.

31 Upper Miocene index diatoms. (*abst*) Pan Am G 56:70 (1931)
36 Pliocene diatoms from the Kettleman Hills, California. (*abst*) G Soc Am, Pr 1935:382 (1936)

Lott, F. S.

36 (and **Hopkins,** G. R.) Natural gas. U S B M, Min Yb 1936:725-748, California 726, 728-729 (1936)

Louderback, George D.

09 Benitoite, its paragenesis and mode of occurrence, with chemical analyses by Blasdale. Cal Univ, Dp G, B 5:331-380 (1909)
26 Period of scarp production in the Great Basin. (*abst*) N Jb 1926, Abt B, Ref 2:197 (1926)
31 (with **Savage,** J. L., **Hinderlider,** M. C., **Berkey,** C. P., and **Williams,** L. A.) Report of consulting board on safety of the proposed Pine Canyon dam, Los Angeles County. Cal Dp Pub Works, Div Water Resources May 1931:22 pp, 8 pls, 1 map (1931)
31a Geologic conditions at the Lafayette dam. (*abst*) G Soc Am, B 42:295 (1931)
33 Geologic report on Fairview damsite on Trinity River. Cal Dp Pub Works, Div Water Resources, B 26:471-478 (1933)
33a (and **Ransome,** F. L.) Geologic report on Kennett, Iron Canyon, and Table Mountain damsites on Sacramento River.

Louderback, George D.—Cont.

Cal Dp Pub Works, Div Water Resources, B 26:431-454 (1933)

34 Notes on the geologic section near Columbia, California: with special reference to the occurrence of fossils in the auriferous gravels. Carnegie Inst, Wash, Pub Paleontology 440: 7-13 (1934)

36 River action in the San Gabriel Mountains, California. (*abst*) G Soc Am, Pr 1935:327 (1936) (*abst*) Pan Am G 63: 305-306 (1935)

Loudermilk, J. D.

35 (and **Woodford,** A. O.) Black iron sulphide in California crystalline limestone. (*abst*) Pan Am G 63:320 (1935) (*abst*) G Soc Am, Pr 1935:342 (1936)

Loughlin, G. F.

33 (and **Behre,** C. H.) Classification of ore deposits. *In* Ore deposits of the western states. Am I M Eng, Lindgren Volume: 17-55, California 44-45 (1933) (*abst*) Rv Geol et Sci conn 14:566 (1934)

Lowy, H.

31 Eine elektrodynamische methode zur esforschung des erdinnern. Physikalische Zs 32:337-345 (1931) (*abst*) Rv Geol et Sci conn 12:412 (1932)

Luther, L. A.

33 San Francisco's Hetch Hetchy project. Compressed Air 38:4157-4162 (1933) (*ann*) An Bib Ec G 6:329 (1934)

Lynton, Edward D.

31 Some results of magnetometer surveys in California. Am As Petroleum G, B 15:1351-1370 (1931) (*ann*) An Bib Ec G 4:356 (1931) (*abst*) G Zentralbl 46:424 (1932) Inst Pet Tech, J 18:41A (1932)

Macelwane, J. B.

33 Tectonic earthquakes. Nat Res Council, B 90:4-8 (1933)

33a Earthquakes—what are they? Sc Mo 36:457-460, California 458 (1933)

MacFarlane, J. M.

31 The quantity and sources of our petroleum supplies. Noel Printing Co. 250 pp (1931) (*abst*) Zs Prak G 41: 199-200 (1933)

MacGinitie, Harry

33 Ecological aspects of the floras of the auriferous gravels. (*abst*) G Soc Am, Pr 1933 1:356 (1934)

34 Tertic floras of Trinity County, California. (*abst*) Pan Am G 62:75-76 (1934) (*abst*) G Soc Am, Pr 1934:390 (1935)

Mackintosh, J. B.

91 Mineralogical notes. Am J Sc (3) 41:438 (1891)

MacIntosh, F. G.

34 Rare gem minerals of America. Oregon Mineralogist 2, no 7:3-4, 30, no 8:5-6, 21 (1934)

34a A trip to Death Valley. Min Soc S Cal 3, no 8:28 (1934)

34b Two rare and beautiful California gems. Oregon Mineralogist 2, no 1:10 (1934)

Maenicke

33 Die Ausbeutung des Searles-Seese in Kalifornien. (*abst*) G Zentralbl, Abt A 47:287 (1932)

Maher, Thomas J.

29 The United States Coast and Geodetic Survey—its work in collecting earthquake reports in the state of California. Seism Soc Am, B 19:77-79 (1929)

Makiyama, Jiro

36 The Meisen Miocene of North Korea. Memoirs of the College of Science, Kyoto Imperial Univ 11, no 4, article 8 (1936)

Mann, L.

30 (**Tryon,** F. G.; and **Rogers,** H. O.) Coal. U S B M, Min Res, Calendar Year 1930:599-773, California 614, 707 (1932)

31 (**Young,** W. H.; **Tryon,** F. G.; **Berquist,** F. E.; and **Bennit,** H. L.) Coal. U S B M, Min Res, Calendar Year 1931: 415-510, California 426, 472 (1933)

32 (and **Tryon,** F. G.) Coal. U S B M, Min Res, Calendar Year 1929:673-858, California 700, 703, 743 (1932)

34 (**Young,** W. H.; **Bennit,** H. L.; and **Tryon,** F. G.) Coal. U S B M, Min Yb, Stat App, Calendar Year 1932:373-454, California 376,423 (1934)... 1933:281-360, California 286, 323 (1935)

Manning, Paul D. V.

31 Sodium salts. Min Ind 39:543-558, California 545, 546, 547, 550 (1931)... 40: 496-508, California 497, 499 (1932)... 41:473-484, California 474, 475, 477, 478 (1933)... 42:530-537, California 530, 533, 534, 536 (1934)

Mansfield, G. R.

32 (and **Boardman,** L.) Nitrate deposits of the United States. U S G S, B 838:Vi, 107 pp, 11 pls, California 23-29 (1932) (*abst*) G Zentralbl, Abt A 48:413-414 (1932) Min Absts 5:164 (1932) (*ann*) An Bib Ec G 5:85 (1933)

34 (and **Lang,** W. B.) The Texas- New Mexico potash deposits. Univ Texas, B

Mansfield, G. R.—Cont.

2, 3401:641-832, California 642, 643, 652, 653, 676 (1934)

Mansfield, Wendell C.

31 Some peculiar spiral fossil forms from California and Mexico. U S Nat Mus, Pr 77, art 13:1-3, 2 pls (1930) Biol Absts 5:15302 (1931)

Marden, J. W.

31 (and **Briggs,** F. H.) Titanium and zirconium. Min Ind 39:602-611, California 603-604 (1931)

Marliave, Chester

31 Geology of the Sacramento River Canyon between Cottonwood Creek and Iron Canyon. Cal Dp Pub Works, Div Water Resources, B 26:463-470 (1933)

33 Geological report on upper Pit River damsites. Cal Dp Pub Works, Div Water Resources, B 41:145-148 (1933)

Marsh, Hallan N.

32 (and **Robinson,** B. H.) Advantages of flowing wells through tubing. *In* Petroleum Dev and tech. Am I M Eng, Tr 1932:301-305 (1932)

Marshall, Kenneth

31 (**Byerly,** P. and **Hester,** J.) The natural periods of vibration of some tall buildings in San Francisco. Seism Soc Am, B 21:268-276, 3 figs (1931) (*abst*) Canada, Dom Obs, Pub, Bib Seism 10:228 (1932)

Marshall, P.

26 Crustal movements and geotectonics in the Pacific region: earthquake, crust-tides, variation of mean sea-level. Third Pan-Pacific Sc Cong, Tokyo, Pr 1:440-443 (1926)

Marshall, W. C.

34 Description of Manhattan Beach area. Cal Oil World 27, no 11:8 (1934)

Martin, Lois T.

30 Foraminifera from the intertidal zone of Monterey Bay, California. Stanford Univ, Micropaleontology B 2, no 3:50-54, figs (1930)

31 Additional notes on the foraminifera from the intertidal zone of Monterey Bay, California. Stanford Univ, Micropaleontology B 3:13-15 (1931)

35 Foraminifera from Monterey Bay. (*abst*) Pan Am G 63:378 (1935) (*abst*) G Soc Am, Pr 1935:416 (1936)

Mason, Herbert L.

31 Pleistocene floras of the San Francisco Bay region. (*abst*) G Soc Am, B

Mason, Herbert L.—Cont.

42:365 (1931) (*abst*) Pan Am G 55:159 (1931)

32 A phylogenetic series of the California closed-cone pine suggested by the fossil record. Madroño, Cal Botanical Soc, 2:49-55 (1932) (*abst*) Pale Zentralbl 5:252 (1934) Biol Absts 7:453, entry 4318 (1933)

33 (and **Chaney,** R. W.) A Pleistocene flora from the asphalt deposits at Carpinteria, California. Carnegie Inst Wash, Pub 415:45-79, 9 pls (1933) Biol Absts 9:3721 (1935) (*abst*) Pale Zentralbl 4:314 (1934)

34 Pleistocene flora of the Tomales formation. Carnegie Inst, Wash, Contr Paleontology (Studies of the Pleistocene paleobotany of California) Pub 415:82-179, 11 tables, 1 fig (1934) (*abst*) N Jb 1935, Ref 3:152-153 (1935)

34a (and **Chaney,** R. W.) A Pleistocene flora from Santa Cruz Island, California. Carnegie Inst, Wash, Pub 415:1-24, 7 pls, 1 fig (1934) (*abst*) N Jb 1931, Ref 3:869 (1931) (*abst*) G Zentralbl, Abt A 48:32 (1932) (*abst*) Pale Zentralbl 1:313 (1932)... 2:143-144 (1932)

Mason, John F.

35 Fauna of the Cambrian Cadiz formation, Marble Mountains, California. S Cal Ac Sc, B 34:97-120 (1935)

36 (and **Hazzard,** J. C.) Middle Cambrian formations of the Providence and Marble Mountains, California. G Soc Am, B 47:229-240, 1 fig (1936)

Matthes, Francois Emile

31 Geologic history of the Yosemite Valley. (*Review*) Geog Rv 21:698-699 (1931) (*abst*) Eng Index 1931:655 (1931) (*Review*) Sierra Club B 17:126-127 (1932) (*abst*) Zs Geom 7:50-53 (1932)

32 Report of the committee on glaciers. Am Geop Union, Tr 1932:282-287, California 284 (1932)... 1933:345-350, California 347 (1933)... 34:279-285, California 281-283 (1934)... 35:387-392, California 390, 391 (1935)

33 Geography and geology of the Sierra Nevada. Int G Cong, Guide Book 16:26-40 (1933) (*abst*) N Jb 1936, Ref 3:67-69 (1936)

33a Up the western slope of the Sierra Nevada by way of the Yosemite Valley. Int G Cong, Guide Book 16:67-81 (1933) (*abst*) N Jb 1936, Ref 3:69-70 (1936)

33b The little "Lost Valley" on Shepherd's Crest. Sierra Club B 18:68-80 (1933)

35 Why we should measure our glaciers. Sierra Club B 20:20-28 (1935)

Matthes, Francois Emile—Cont.

35a Studies in glacial sediments. Nat Res Council, B 98:82-146, California 139-141 (1935)

Matthew, William Diller

31 A Pliocene mastodon skull from California. *Pliomastodon vexillarius* n sp. (*abst*) N Jb 1931, Ref 3:682 (1931) G Zentralbl 45:458 (1931) (*abst*) Pale Zentralbl 2:328 (1933)

31a (and **Stirton,** R. A.) Osteology and affinities of Borophagus. (*abst*) N Jb 1931, Ref 3:678-679 (1931) (*abst*) G Zentralbl 45:458 (1931) (*abst*) Pale Zentralbl 1:142 (1932)

Maxson, John H.

31 Geomorphic features of northwesternmost California. (*abst*) G Soc Am, B 43:224 (1932) (*abst*) Pan Am G 55:358-359 (1931)

31a Tertic mammalian fauna from Mint Canyon formation of southern California. (*abst*) G Zentralbl 44:66-67 (1931)

32 Structural relationships of coast and continental margin of northern California. (*abst*) Pan Am G 58:66-67 (1932) (*abst*) G Soc Am, B 44:152 (1933)

33 Contact conditions of some chromite deposits in serpentine in the Klamath Mountains. (*abst*) G Soc Am, B 44:166 (1933) (*abst*) Pan Am G 58:77-78 (1932)

33a Economic geology of portions of Del Norte and Siskiyou Counties, northwesternmost California. Cal Dp Nat Res, Div Mines, M G, J, St Mineralogist's Rp 29:123-160 (1933) (*abst*) Eng Index 1934: 710 (1934)

34 Pre-Cambrian stratigraphy of the Inyo Range. (*abst*) Pan Am G 61:310-311 (1934)

34a (and **Anderson,** G. H.) Physiography of northern Inyo Range. (*abst*) G Soc Am, Pr 1934:318 (1935) (*abst*) Pan Am G 61:314-315 (1934)

35 (and **Davis,** W. M.) Valleys of the Panamint Mountains, California. (*abst*) G Soc Am, Pr 1934:339 (1935)

35a (and **Anderson,** G. H.) Terminology of surface forms of the erosion cycle. J G 43:88-96 (1935)

Mayo, Evans B.

30 Preliminary report on the geology of southwestern Mono County, California. Cal Dp Nat Res, Div Mines, Mining in California, St Mineralogist's Rp 26:475-482, 3 figs (1930)

31 Fossils from the eastern flank of the Sierra Nevada, California. Science 74: 514-515 (1931) (*abst*) Pale Zentralbl 3:312 (1933) Biol Absts 7:454, entry 4326 (1933)

Mayo, Evans B.—Cont.

32 Two new occurrences of piedmontite in California. Am Mineralogist 17: 238-248, 3 figs (1932) (*abst*) Rv Geol et Sci conn 13:264 (1933) Am Mineralogist 17:117 (1932) Min Absts 5:222 (1933) Chem Absts 26:4773 (1932)

33 Discovery of piedmontite in the Sierra Nevada. Cal Dp Nat Res, Div Mines, M G, J, St Mineralogist's Rp 29: 239-243 (1933)

33a Preliminary survey of an intraseptum intrusion in eastern California. (*abst*) G Soc Am, Pr 1933:97 (1934)

34 Geology and mineral deposits of Laurel and Convict Basins, southwestern Mono County, California. Cal Dp Nat Res, Div Mines, M G, J, St Mineralogist's Rp 30:79-90 (1934)

34a The Pleistocene Long Valley Lake in eastern California. Science, n s 80:95-96 (1934)

35 Some intrusions and their wall rocks in the Sierra Nevada. J G 43:673-789 (1935) (*abst*) Sc Progress 31:316 (1936)

36 (**Conant,** L. C., and **Chelikowsky,** J. R.) Southern extension of the Mono Craters, California. Am J Sc 32 (5):81-97, figs, maps (1936)

McCollough, E. H.

34 Structural influence of the accumulation of petroleum in California. *In* Problems of petroleum geology. Am As Petroleum G, Sidney Powers Memorial Volume: 735-760 (1934) (*ann*) An Bib Ec G 7:314 (1934)

McComb, H. E.

36 Observations on recent progress in seismology. (*abst*) Pan Am G 65:157 (1936)

McDonald, G. A.

34 (and **Shepard,** F. P.) Sediments of Santa Monica Bay. (*abst*) Pan Am G 61: 317 (1934)

McDowell, G.

30 (Jensen, J.; Goold, W. D.; and Gwin, M. L.) Deep sand development at Santa Fe Springs. *In* Petroleum Dev and Tech Am I M Eng, Tr 1930:310-321 (1930)

McGlashan, H. D.

31 (**Grover,** N. C., **Henshaw,** F. F., and **Canfield,** G. H.) Surface water supply of the United States 1927 Part XI Pacific Slope basins in California. U S G S, W-S P 1927, 651:IX, 299 pp (1931)... 1928, 671:IX, 304 pp (1931)... 1929, 691:IX, 294 pp (1931)... 1930:706:IX, 317 pp (1932)... 1931, 721:XI, 497 pp (1932)...

McGlashan, H. D.—Cont.

1932, 736:XI, 415 pp (1933)... 1933, 751:XI, 376 pp (1935)

McKenzie, M. R.

32 (and **Lancaster, H. K.**) Milling methods at the Concentrator of the Walker Mining Co., Walkermine, California. U S B M, Inf Circ 6555:1-11 (1932)

McMasters, John H.

32 Eocene Llajas formation of California. (*abst*) Pan Am G 58:148-149 (1932) (*abst*) G Soc Am, B 44:217 (1933)

33 Notes on a Middle Eocene formation of Ventura County, California. Stanford Univ, Micropaleontology B 4:64-71 (1933) (*abst*) Pale Zentralbl 5:213 (1934)

Mead, Roy G.

31 A brief history of Kettleman Hills. Oil B 17:507-511, 597-602, 645 (1931)

33 Kramer borax deposit in California and the development of other borate ores. M Metal 14:405-409 (1933) Chem Absts 28: 6660 (1934)

Means, T. H.

32 Death Valley. Sierra Club B 17:67-76 (1932)

Mehl, Maurice G.

30 Petroliferous provinces. Science 1930: 541-543 (1930)

Mejea, J.

35 (**Jomercq, J. Junr.**, and **Uren, L. C.**) Large diameter wells are indicated. Petroleum World (London) 32:260-262 (1935)

Melhase, John

33 Diatomaceous earth, its nature, occurrence and use. Rocks & Min 8:27-29 (1933)

34 A diversity of California minerals. Oregon Mineralogist 2, no 7:7-8, 23 (1934)

35 Fluorescent minerals of California. The Mineralogist 3, no 1:3-4, 38 (1935)

35a Discovery of sanbornite in California. The Mineralogist 3, no 9:3-4, et seq (1935)

35b Some garnet localities of California. The Mineralogist 3, no 11:7-8, 22-24 (1935)

36 A new occurrence of rare-earth minerals in California. The Mineralogist 4, no 1:11 (1936)

36a Industrial uses of nonmetallic minerals. The Mineralogist 4, no 8:7-8 (1936)

Melton, Frank Armon

34 Erosional soil hillocks. (*abst*) G Soc Am, Pr 1933:98 (1934)

Melville, William Harlow

92 Tourmaline from Nevada County, California. U S G S, B 90:39 (1892)

Mendenhall, W. C.

31 Fifty-second annual report of the Director of the United States Geological Survey to the Secretary of the Interior for the fiscal year ending June 30, 1931. U S Dp Interior, An Rp 1931:95 pp (1931)... Fifty-third, 1932:94 pp (1932)... Fifty-fourth, 1933:203-237 (1933)... Fifty-fifth, 1934:217-253 (1934)... Fifty-sixth, 1935:233-274 (1935)

Menken, F. A.

32 (with **Leach, C. E.**) Overturned plunge on overturned folds in Sespe-Piru Creek district, California. Am As Petroleum G, B 16:209-212 (1932) (*ann*) An Bib Ec G 5:140 (1932) (*abst*) Eng Index 1932:613 (1932)

Merriam, Charles Warren

31 Notes on a brittle star limestone from the Miocene of California. Am J Sc (5) 21:304-310, 2 figs (1931) (*abst*) Pale Zentralbl 1:227 (1932)... 2:298 (1933) (*abst*) N Jb 1935, Ref 3:467 (1935)

33 Zonal distribution and foreign affinities of Turritella occurring in Cretaceous, Tertiary and Quaternary deposits on the Pacific Coast of North America. (*abst*) G Soc Am, B 44:216 (1933) (*abst*) Pan Am G 58:147-148 (1932) Pale Zentralbl 4:276 (1934)

34 Middle Eocene faunas of northern California. (*abst*) Pan Am G 62:77-78 (1934) (*abst*) G Soc Am, Pr 1934:392 (1935)

Merriam, John Campbell

30 The living past. 144 pp (1930) (*Review*) Sc Progress 25:336 (1930)

31 The cats of Rancho La Brea: a climax in evolution. (*abst*) Science 74:576 (1931)

32 (and **Stock, C.**) The Felidae of Rancho La Brea. Carnegie Inst, Wash, Pub 422, 152 figs, 42 pls (1932) (*abst*) N Jb 1934, Ref 3:319-320 (1934) Biol Absts 9:491, entry 4313 (1935)

33 (and **Stock, C.**) Tertiary mammals from the auriferous gravels near Columbia, California. Carnegie Inst, Wash, Pub Paleontology 440:1-6, 2 figs (1934)

Merriam, R.

36 (and **Laudermilk, J. D.**) Two diopsides from southern California. Am Mineralogist 21:715-718 (1936)

Merrill, Charles White

32 Tin. U S B M, Min Res, Calendar Year 1929:333-362, California 337 (1932)

34 (and Heikes, V. C.) Gold, silver, copper, lead, and zinc in California. U S B M, Min Yb, Stat App 1932-1933:199-215 (1934)

36 (and Gaylord, H. M.) Gold, silver, copper, lead, and zinc in California. U S B M, Min Yb 1936:219-241 (1936)

Metcalf, R. W.

32 Clay. U S B M, Min Yb, Stat App 1932-1933:1-5, California 2 (1934)

32a (and Galiher, C.) Gypsum. U S B M, Min Yb, Stat App 1932-1933:17-23, California 18, 20 (1934)-

34 (and Rogers, H. O.) Feldspar. U S B M, Min Yb 1934:999-1007, California 1005, 1006, 1007 (1934)... 1935:1107-1114. California 1112 (1935)

34a (and Tyler, P. M.) Clay. U S B M, Min Yb 1934:873-887, California 876 (1934)... 1935:977-993, California 980 (1935)

34b (and Adams, W. W.) Fuller's earth. U S B M, Min Yb 1934:969-974, California 971, 974 (1934)... 1935:1063-1068 (1935)

35 (and Tyler, P. M.) Gypsum. U S B M, Min Yb 1935:949-966, California 951, 953 (1935)

36 Feldspar. U S B M, Min Yb 1936: 981-987 (1936)

36a (and Furness, J. W.) Copper. U S B M, Min Yb 1936:107-135, California 112, 114, 116-118 (1936)

36b Mercury. U S B M, Min Yb 1936: 413-424, California 414, 417, 418, 420 (1936)

36c (and Geyer, L. E.) Clay. U S B M, Min Yb 1936:867-875, California 869, 871 (1936)

Meyer, Helena M.

29 (with Julihn, C. E.) Copper. U S B M, Min Res, Calendar Year 1929:525-580, California 527, 538, 539, 540, 541, 542, 543, 544, 545, 557-558 (1932)... 1930:691-748, California 725 (1933)... 1931:575-600, California 582, 583, 584, 585, 586, 587, 590, (1934)... Min Yb 1932-1933:27-52, California 30, 32, 33, 34, 35, 36, 37 (1933)... 1934: 53-77, California 63-64 (1934)

31 (and Tyler, P. M.) Mercury. U S B M, Min Res, Calendar Year 1931:191-209, California 198-201 (1934)

32 (and Jackson, C. F.) Mercury. U S B M, Min Res, Min Yb 1932-1933:225-241, California 229, 230, 231-233 (1933)... 1934: 385-397, California 385, 390-391, 393 (1934)

32a (and Wright, C. W.) Lead. U S B M, Min Yb 1932-1933:53-66, California 58 (1933)

33 (and Pehrson, E. W.) Lead and zinc pigments and zinc salts. U S B M, Min

Meyer, Helena M.—Cont.

Yb 1932-1933:87-90, California 95 (1933)... 1934:123-137, California 132, 133 (1934)... 1935:123-135, California 132 (1935)

34 (and Pehrson, E. W.) Lead. U S B M, Min Yb 1934:79-99, California 86-87, (1934)... 1935:75-98, California 85 (1935)... 1936:137-156, California 143 (1936)

35 (and Furness, J. W.) Mercury. U S B M, Min Yb 1935:437-463, California 442, 449-451, 458 (1935)

35a (Furness, J. W.; and Pehrson, E. W.) Copper. U S B M, Min Yb 1935:45-73, California 51, 53, 54, 55, 56 (1935)

36 (and Furness, J. W.) Copper. U S B M, Min Yb 1936:107-135, California 112, 114, 116-118 (1936)

Middleton, Jefferson

32 Fuller's earth. U S B M, Min Res, Calendar Year 1929:61-64, California 61 (1932)... 1930:69-71, California 69 (1932)

32a Clay. U S B M, Min Res, Calendar Year 1929:95-104, California 98, 100, 102 (1932)

32b (and Bowles, O.) Feldspar. U S B M, Min Res, Calendar Year 1929:83-93, California 89 (1932)

32c (and Santmyers, R. M.) Gypsum. U S B M, Min Res, Calendar Year 1929: 105-118, California 106 (1932)... 1930:87-100, California 90, 91, 92, 93, 94, 95 (1932)... 1931:191-203, California 193, 195, 199, (1933)

32d (and Herlihy, K. V.) Clay. U S B M, Min Res, Calendar Year 1930:171-180, California 174, 176, 178 (1932)

32e (and Hughes, H. H.) Feldspar. U S B M, Min Res, Calendar Year 1930: 137-149, California 143, 145-146 (1932)... 1931:179-190, California 184, 186, 187 (1933)

33 (and Hatmaker, P.) Fuller's earth. U S B M, Min Res, Calendar Year 1931: 73-98, California 84, 91 (1933)

Miller, Alden Holmes

29 Additions to the Rancho La Brea avifauna. Condor 31:223-224 (1929)

32 The fossil passerine birds from the Pleistocene of Carpinteria, California. Cal Univ Dp G Sc, B 21:169-194, 3 pls (1932) (abst) Pale Zentralbl 1:378 (1932)

33 The passerine remains from the Pleistocene of Carpinteria, California. (abst) N Jb 1933, Ref 3:208 (1933)

33a The passerine remains from Rancho La Brea in the paleontological collections of the University of California. (abst) N Jb 1933, Ref 3:208 (1933) (abst) Pale Zentralbl 1:69 (1932)

34 (and Ashley, J. F.) Goose footprints on a Pliocene mud flat. Condor 36:178-179, 1 fig (1934) (abst) U S G S, B 869: 165 (1933 and 1934)

Miller, Benjamin L.

34 Graphite. Min Ind 42:293-301, California 298 (1934)... 43:295-304, California 298-299 (1935)

Miller, Franklin S.

35 (and **Larsen,** E. S.) The Rosiwal method and the modal determination of rocks. Am Mineralogist 20:260-273, 7 tables (1935)

35a Anorthite from California. Am Mineralogist 20:139-146, 2 figs (1935) Min Absts 6:118 (1935) Chem Absts 29:6538 (1936)

Miller, Fred M.

31 Gold production of Nevada County, California. Cal M G, J 1, no 3:11, 17 (1931)

34 Prosperity rules in Grass Valley and Nevada City. Eng M J 135:511-515 (1934)

Miller, H. C.

32 Migration of injected gas through oil and gas sands of California. U S B M, Rp Invest 3177:29 pp, 9 figs (1932) (*abst*) U S B M, List Pub 1910-1932:121 (1933) (*ann*) An Bib Ec G 5:136 (1932)

Miller, Harold W.

33 (and **Wilhelm,** V. H.) Developments in the California petroleum industry during 1932. *In* Petroleum Dev and Tech, Am I M Eng, Tr 1933:345-351 (1933)... 1933, Am I M Eng, Tr 1934:182-197 (1934)

34 (and **Wilhelm,** V. H.) Developments in the California oil industry during the year 1933. Am I M Eng, Tr 107:182-197 (1934)

Miller, Loye Holmes

31 Bird remains from the Kern River Pliocene of California. Condor 33:70-72, 1 fig (1931) (*abst*) Pale Zentralbl 1:377 (1932)

31a Pleistocene birds from the Carpinteria asphalt of California. Cal Univ Dp G Sc, B 20:361-374, 4 figs (1931) (*abst*) N Jb 1933, Ref 3:208 (1933) (*abst*) Pale Zentralbl 1:377-378 (1932)

32 Two Pleistocene avifaunas from the Carpinteria asphalt. (*abst*) G Soc Am, B 43:291 (1932)

32a The Pleistocene storks of California. Condor 34:212-216, 3 figs (1932) (*abst*) Pale Zentralbl 3:149 (1933)

32b Royal vulture from the Kern River Pliocene, California. (*abst*) G Soc Am, B 43:291-292 (1932) (*abst*) Pale Zentralbl 2:321 (1933)

32c Further bird remains from the upper San Pedro Pleistocene. (*abst*) Pale Zentralbl 1:377 (1932)

Miller, Loye Holmes—Cont.

32d A fossil goose from the Ricardo Pliocene. (*abst*) Pale Zentralbl 2:264 (1933) (*abst*) N Jb 1933, Ref 3:207 (1933)

33 The Lucas auk of California. Condor 35:34-35 (1933) (*abst*) Pale Zentralbl 3:274 (1933)

33a A Pleistocene record of the flammeolated screech owl. San Diego Soc N H, Tr 7:209-210 (1933) (*abst*) Pale Zentralbl 4:178 (1934)

34 A new horizon for the extinct goose *"Chendytes."* Science n s 80:141-142 (1934)

35 A second avifauna from the McKittrick Pleistocene. Condor 37:72-79 (1935)

35a New bird horizons in California. Cal Univ, Los Angeles Biol Sc, Pub 1, no 5:73-80 (1935)

Miller, R. B.

36 Nickel. U S B M, Min Yb 1936:521-524, California 521 (1936)

Miller, R. H.

31 Analysis of some torsion balance results in California. Am As Petroleum G, B 15:1419-1429 (1931) (*ann*) An Bib Ec G 4:353 (1931) (*abst*) G Zentralbl 46:424 (1932) Inst Pet Tech, J 18:42A (1932)

Miller, William John

30 Geomorphology of the southwestern San Gabriel Mountains of California. (*abst*) Petermann's Mitt 76:273 (1930)

31 The landslide at Point Firmin, California. Sc Mo 32:464-469, 5 figs (1931) (*abst*) Canada, Dom Obs, Pub Bib Seism 10:273 (1932)

31a Anorthosite in Los Angeles County, California. J G 39:331-344, 1 fig (1931) (*abst*) Rv Geol et Sci conn 12:507 (1932) (*abst*) G Zentralbl 45:211 (1931) (*Review*) Sc Progress 27:439 (1933) (*abst*) N Jb 1932, Band 2, Ref 2:878 (1932)

31b Geologic section across the southern Sierra Nevada of California. Cal Univ Dp G Sc, B 20:331-360, 3 figs, 4 pls (1931) (*abst*) G Zentralbl, Abt A 48:508 (1933)

31c Geology of Deep Spring Valley, California. (*abst*) N Jb 1931, Ref 3:557-558 (1931)

32 Cenozoic history of San Gabriel Mountains. (*abst*) Pan Am G 58:78 (1932) (*abst*) G Soc Am, B 44:166-167 (1933)

32a Comparison of two granites, San Gabriel Mountains, California. (*abst*) Pan Am G 58:74 (1932) (*abst*) G Soc Am, B 44:161-162 (1933)

32b Intrusive power of anorthosite. (*abst*) Pan Am G 57:233 (1932)

Miller, William John—Cont.

35 Geology of the western San Gabriel Mountains of California. Cal Univ, Los Angeles, Pub Math and Phys Sc 1:1-114, 13 pls, 6 figs, 1 map (1935) (*abst*) Am Mineralogist 20:402-403 (1935) (*abst*) Rv Geol et Sci conn 15:296 (1935)

35a Pre-Cambrian and associated rocks near Twenty-nine Palms, California. (*abst*) G Soc Am, Pr 1934:99 (1935)

35b A geologic section across the southern Peninsular Range of California. Cal Dp Nat Res, Div Mines, M G, J, St Mineralogist's Rp 31:115-143, map (1935) (*abst*) G Soc Am, Pr 1933:312 (1934) (*ann*) An Bib Ec G 8:256 (1936) (*abst*) G Zentralbl, Abt A, Bd 58:233-234 (1936)

35c Geomorphology of the southern Peninsular Range of California. G Soc Am, B 46:1535-1582, 5 pls, 1 fig (1935) (*abst*) G Soc Am, Pr 1934:317-318 (1935) (*abst*) Pan Am G 61:313 (1934)

Millett, E. R., Jr.

35 New Playa Del Rey area may be another major field for Los Angeles Basin. Petroleum World 1935 (April): 13-15, 30 (1935)

Mills, B.

31 Everything is unusual at Kettleman Hills. Oil Weekly 60, no 5:37-44, 10 figs (1931)

31a Ventura probably has oil below 15,-000 feet. Oil Weekly 62, no 12:18-21 (1931) (*ann*) An Bib Ec G 4:317 (1931) (*ann*) An Bib Ec G 4:317 (1931) (*abst*) Eng Index 1931:1036 (1931)

32 Geology, both good and bad, has figured in California development. Oil Weekly 65:28-32 (1932)

33 Two holes under one derrick offer an economy when foundations are expensive. Oil Weekly 71:14-16, no 12 (1933) (*abst*) Rv Geol et Sci conn 14:357 (1933-1934)

34 Geophysical work hindered by geologic conditions in California. Oil Weekly 73, no 10:22-23 (1934) (*abst*) Inst Pet Tech, J 20:389A (1934) (*ann*) An Bib Ec G 7:121 (1935) (*abst*) Eng Index 1934: 520 (1934)

34a California crude reserves are below future demand requirements. Oil Weekly 76, no 1:13-16 (1934-1935)

35 Kettleman Hills today—water intrusions and declining pressures lower estimated ultimate production. Oil Weekly 80, no 1:24-30 (1935) (*ann*) An Bib Ec G 8:358 (1936)

35a Geophysical operations have been very successful in San Joaquin valley during past year. Oil Weekly 78, no 4: 28-29 (1935) (*ann*) An Bib Ec G 8:400

Mills, B.—Cont.

(1936) (*abst*) Rv de Geol et Sci conn 16: 180 (1936)

The Mineral Collector

07 Tripoli deposits in California. Mineral Collector 14, no 6:95-96 (1907)

Mineral Industry

32 Magnesite. Min Ind (Roush, G. A., editor) 40:351-359, California 351, 352 (1932)... 41:343-349, California 343, 344 (1933)... 42:380-386, California 380, 382 (1934)

33 Quicksilver. Min Ind (Roush, G. A., editor) 41:455-461, California 455, 456 (1933)... 42:510-518, California 513 (1934)... 43:509-515, California 511-512 (1935)

33a Zinc. Min Ind (Roush, G. A., editor) 41:536-558, California 543 (1933)... 42:582-613, California 595 (1934)... 43:584-615, California 594 (1935)

33b Lead. Min Ind (Roush, G. A., editor) 41:321-342, California 323, 328, 329 (1933)... 42:358-379, California 362, 364 (1934)

34 Gypsum. Min Ind (Roush, G. A., editor) 42:303-307, California 304 (1934)

34a Talc and soapstone. Min Ind (Roush, G. A., editor) 42:549-552, California 549, 550 (1934)... 43:549-552, California 549 (1935)

35 Iodine. Min Ind (Roush, G. A., editor) 43:634-636, California 635 (1935)

Mining and Scientific Press

80 Theory of auriferous gravel channels. M Sc Press 41:226 (1880)

Mining Journal, Arizona

31 Important ore discovery at Carson Hill, M J Ariz 15, no 11:22, 42 (1931)

31a Unique placer development (Yuba River). M J Ariz 14, no 23:4 (1931)

Miser, H. D.

20 (and **Fairchild**, J. G.) Hausmannite in the Batesville district, Arkansas. Wash Ac Sc J, 10:1-8 (1920)

Mitchell, A. W.

32 (and **Ridgway**, R. H.) Sulphur and pyrites. U S B M, Min Yb 1932-1933:669-686, California 671, 677, 681 (1933) ... 1934:907-928, California 918-919, 924 (1934)... 1935:1011-1028, California 1020, 1024 (1935)

36 (and **Ridgway**, R. H.) Pyrites. U S B M, Min Yb 1936:909 (1936)

Mitchell, R. L.

34 (and **Ritter**, G. J.) Composition of three fossil woods mined from the Miocene auriferous gravels of California.

Mitchell, R. L.—Cont.

Am Chem Soc, J 56:1603-1605, 1 fig July (1934) Chem Absts 28:5376 (1934)

Mitchell, Stewart

33 A 320-foot concrete arch on scenic route along California coast. Eng News-Rec 110 no 15:467-470 (1933) (ann) An Bib Ec G 6:324 (1934)

Miyabe, Naomi

30 Deformation of earthcrust in California. Earthquake Res Inst, B 8, no 1:45-59, 10 figs (1930) (abst) Canada, Dom Obs, Pub, Bib Seism 10:108 (1930) (abst) G Zentralbl, Abt A 46:765 (1932)

Moehlman, Robert S.

34 (and Gonyer, F. A.) Monticellite from Crestmore, California. Am Mineralogist 19:474-476 (1934) (abst) N Jb 1935, Ref 1:62 (1935) Min Absts 6:96 (1935)

Moodie, Roy Lee

29 Studies in paleodontology. 20 The teeth and jaws of "Nothrotherium." Pacific Dental Gazette 37:677-680, 3 figs (1929)

29a Studies in paleodontology. 16 The California sabre-tooth: the mandibular teeth and asssociated structures. Pacific Dental Gazette 37:317-321, 6 figs (1929)

30 Studies in paleopathology. 28 The phenomenon of sacralization in the Pleistocene sabre-tooth. Am J Surgery 10:587-589 (1930)

Moody, Graham B.

35 Unconformity exposed in Santa Ana Mountain foothills. (abst) Am As Petroleum G, B 19:1841-1842 (1935)

35a Geology of Santa Rosa Island. (abst) Pan Am G 63:316-317 (1935) (abst) Am As Petroleum G, B 19:136 (1935) (abst) G Soc Am, Pr 1935:338 (1936)

Moore, Bernard N.

30 Geology of a portion of the Santa Ana Mountains. Orange County, California. (abst) G Soc Am, B 42:291-292 (1931) (abst) Pan Am G 54:69 (1930)

35 Some strontium deposits of southeastern California and western Arizona. Am I M Eng (T P 599 Class H Nonmetallic Minerals 25:Class I, Mining Geology 53) (abst) M Metal 16:115 (1935) (ann) An Bib Ec G 8:94 (1935) Chem Absts 29:3944 (1935) (abst) G Zentralbl, Abt A, Bd 57:341 (1936)

Moore, Gideon E.

70 Ueber das Vorkommen des amorphen Quecksilbersulfids in der Natur. Journ. f prakt Chem 110:319-329 (1870)

Moorehead, W. R.

31 Methods and costs of mining quicksilver at the New Idria Mine, San Benito County, California. U S B M, Inf Circ 6462:14 pp, 8 figs (1931) (abst) U S B M, List Pub 1910-1932:148 (1933) (abst) Eng Index 1931:865 (1931)

Morris, S. B.

33 Pasadena builds Pine Canyon dam. Civil Eng 3:309-313 (1933) (ann) An Bib Ec G 6:322 (1934) (abst) G Zentralbl, Abt A, Bd 53:486 (1934)

Morse, Roy R.

35 (and Bailey, T. L.) Geological observations in the Petaluma district, California. G Soc Am, B 46:1437-1456, 1 pl, 2 figs (1935)

Moyer, Dorothy A.

29 Shallow water foraminifera from off Point Firmin, San Pedro, California. Stanford Univ. Micropaleontology B 1, no 11:5-7 (1929)

Muller, Siemon

31 (Kerr, P. F.; and Schenck, H. G.) Geology of the Ventura quadrangle, California. (abst) G Soc Am, B 42:186-187 (1931) (abst) Pan Am G 55:64 (1931)

35 North American upper Triassic Pelecypod. *Pseudomonotis subcircularis* (Gabb). (abst) G Soc Am, Pr 1934:388 (1935)

Mullerried, Ederico G.

32 Monografia del genero Coralliochama. Am Inst Biol 3:169-179 (1932) (abst) Rv Geol et Sci conn 13:194 (1933)

Mulryan, Henry

36 Geology, mining, and processing of diatomite at Lompoc. Santa Barbara County, California. Am I M Eng, Tech Pub 687:30 pp (1936)

Murdoch, Joseph

34 Amber in California. J G 42:309-310 (1934) (ann) An Bib Ec G 7:66 (1935) (abst) N Jb 1935, Ref 1:364 (1935) Chem Absts 28:6092 (1934) (abst) G Zentralbl, Abt A, Bd 53:146-147 (1934)

35 Quartz-fluorite pseudomorphs. (abst) Pan Am G 64:69-70 (1935) (abst) G Soc Am, Pr 1935:346-347 (1936)

36 Silica-fluorite pseudomorphs. Am Mineralogist 21:18-32 (1936) (abst) Rv Geol et Sci conn 16:491 (1936)

36a Andalusite in pegmatite. Am Mineralogist 21:68-69 (1936)

36b Adamite from Chloride Cliff, California. Am Mineralogist 21:811-813 (1936)

Murphy, F. Mac

30 Geology of the Panamint silver district, California. (*abst*) Rv Geol et Sci conn 11:438 (1930) (*abst*) G Zentralbl 43: 407-408 (1930-1931) (*abst*) Zs Prak G 39: 48 (1931) Chem Absts 25:2669 (1931)

31 Geology and ore deposits of a part of the Panamint Range, Inyo County, California. (*abst*) N Jb 1931, Ref 2:447 (1931)

31a Dumortierite in Riverside County, California. (*abst*) N Jb 1931, Ref 2:447 (1931)

32 Geology of a part of the Panamint Range, California. Cal Dp Nat Res, Div Mines, Mining in California, St Mineralogist's Rp 28:329-356 (1932)

Murphy, P. C.

30 (and **Judson, S. A.**) Deep sand development in Barbers Hill. Am As Petroleum G, B 14:719-741, 11 figs (1930) Oil Gas J 28, no 44:146-147 (1930) Oil Weekly 57, no 3:25-30, 76, 78 (1930) (*abst*) Eng Index 1930:1300 (1930)

Musser, E. H.

31 Buttonwillow gas field. (*abst*) Eng Index 1931:948 (1931) (*ann*) An Bib Ec G 4:101 (1931)

31a Operations in district No. 4, 1930. Cal Dp Nat Res, Div Oil and Gas, California Oil Fields 16, no 3:52-65 (1931)... 1931, 17, no 3:41-49 (1932)... 1932, 18, no 3:40-48 (1933)... 1933, 19, no 3:38-48 (1934)... 1934, 20, no 3:38-48 (1935)

Myers, D. B.

30 Paleontology, an aid to petroleum geology. Union Oil B 11, no 5:1-4, 4 figs. (1930) (*abst*) Eng Index 1930:1300 (1930)

Narici, Enrico

33 Il problema dell'energia. L'utilizzazione delle forze vulcaniche. L'Industria Mineraria, 5:411-417 (*abst*) Zs Vulkan 14:324 (1933)

National Petroleum News.

33 Earthquake causes temporary shutdown at Signal Hill. Damage small. Nat Pet News 25, no 11:23-24 (1933)

36 The drill ever deeper in the search for crude. Nat Pet News 28, no 6:256 (1936)

Natland, Manley L.

33 Distribution of some recent and fossil foraminifera in southern California. (*abst*) Pan Am G 59:378-379 (1933) (*abst*) G Soc Am, Pr 1933 1:393 (1934)

33a The temperature and depth distribution of some recent and fossil foraminifera in the southern California region.

Natland, Manley L.—Cont.

Scripps Inst Oceanography, B 3:225-230 (1933)

Nature

31 The La Brea asphalt pits. Nature 127:69 (1931)

33 The California earthquake of March 10. Nature 131:391, 409-510 (1933)

Nelson, Richard N.

31 (and **Schenck, H. G.**) Eocene algae and stellate orbitoids from the Santa Ynez Range, California. (*abst*) G Soc Am, B 42:371 (1931) (*abst*) Pan Am G 54:240 (1930)

Nelson, W. I.

30 Mining methods and costs at the Engels mine, Plumas County. U S B M, Inf Circ 6260:22 pp (1930) M Cong J 16 no 9:673-705, 24 figs (1930) (*abst*) Eng Index 1930:454 (1930)

31 Milling methods and costs at the Superior Concentrator of the Engels Copper Mining Company, Plumas County, Calif. U S B M, Inf Circ 6550:1-23 (1931)

Neumann, Frank

29 The velocity of seismic surface waves over Pacific Paths. Seism Soc Am, B 19:63-76 (1929)

32 (and **Bodle, R. R.**) United States earthquakes, 1930. U S Coast Serial 539: 25 pp (1932) (*abst*) Canada, Dom Obs, Pub, Bib Seism 10:274 (1932)

33 The strong-motion records of the southern California earthquake of March 10, 1933. (*abst*) Seism Soc Am, Eastern Sec 5:272 (1933) (*abst*) Am Geop Union, Tr 14:272 (1933) (*abst*) Wash Ac Sc, H 23:536 (1933)

33a (and **Heck, N. H.**) Destructive earthquake motions measured for first time. Eng News-Rec 110:804-807, 8 figs, 1 table (1933) (*abst*) Canada, Dom Obs, Pub, Bib Seism 10:345 (1933)

Newitt, Harry R.

36 Future possibilities of prospecting. The Pacific Mineralogist 3, no 2:16-17 (1936)

Newman, William A.

36 Some practical points on hydraulic mining. Eng M J 137:10-12 (1936)

Newsom, John F.

03 Clastic dykes. G Soc Am, B 14:227-268, pls 21-31, figs 1-19 (1903) (*abst*) N Jb Band 1, 1906:224-225 (1906)

Nickell, F. A.

30 Geology of Soledad quadrangle. (*abst*) Pan Am G 54:157 (1930) (*abst*) G Soc Am, B 42:313-314 (1931)

Nickles, J. M.

31 Bibliography of North American geology, 1919-1928. U S G S, B 823:1005 pp (1931)... 1929-1930 U S G S, B 834:ii, 280 pp (1931)... 1931-1932. U S G S, B 858: 300 pp (1934) (*abst*) Rv Geol et Sci conn 14:398 (1934)

Noble, E. B.

32 Probing our geological past. Union Oil Company, B 13:6-10, 15 (1932)

Noble, Levi F.

31 Nitrate deposits in southeastern California, with notes on deposits in southeastern Arizona and southwestern New Mexico. U S G S, B 820:108 pp, 7 figs, 19 pls (1931) (*abst*) Rv Geol et Sci conn 13:500 (1933) (*ann*) An Bib Ec G 4:72 (1931) (*abst*) G Zentralbl, Abt A, 47:159 (1932) Nature 128:155 (1931) (*abst*) N Jb 1932, Band 2, Ref 2:58 (1932) (*abst*) Eng Index 1931:961 (1931) Chem Absts 25:3597 (1931)

32 Excursion to the San Andreas fault and Cajon Pass. Int G Cong, Guide Book 15:10-21 (1932)

32a The San Andreas rift and some other active faults in the desert region of southeastern California. Carnegie Inst, Wash, Yb 31:355-363 (1932) Third Pan-Pacific Sci Cong, Tokyo, Pr:394-401 (1928) (*abst*) N Jb 1931, Ref 2:305 (1931)

34 Rock formations of Death Valley, California. Science 80:173-178 (1934)

Nolan, T. B.

33 Epithermal precious-metal deposits. *In* Ore deposits of the western states. Am I M Eng, Lindgren Volume, Pub: 623-641, California 630 (1933)

33a Mother Lode district. *In* Ore deposits of the western states. Am I M Eng, Lindgren Volume:579 (1933)

Nomland, Jorgen O.

32 (and **Schenck**, H. G.) Cretaceous beds at Slate's Hot Springs, California. Cal Univ, Dp G Sc, B 21:37-49, 4 figs (1932) (*abst*) Rv Geol et Sci conn 12:592 (1932) (*abst*) G Zentralbl, Abt A 48:397-398 (1932) (*abst*) Eng Index 1932:612 (1932)

Norris, A.

34 (and **Franz**, S. I.) Human reactions to the Long Beach earthquake. Seism Soc Am, B 24:109-114 (1934)

Norris, Byron B.

30 Report on the oil fields on or adjacent to the Whittier Fault. Cal Dp Nat Res, Div Gas and Oil, Cal Oil Fields 15, no 4:5-20, 7 maps on supp plates (1930) (*ann*) An Bib Ec G 4:101 (1931)

Oakeshott, G. B.

34 (and **Clements**, T.) Eocene (Martinez) of San Gabriel Mountains. (*abst*) Pan Am G 61:307-308 (1934) (*abst*) G Soc Am, Pr 1934:310 (1935)

O'Farrell, Charles

33 Microscopic examination of chromite ore from Klamath River District, California. (*abst*) Utah Ac Sc, Pr 10:69 (1933)

Oil and Gas Journal

35 Consistent gain in use of geophysical methods shown in California fields. Oil Gas J 34, no 25:158-162 (1935) (*abst*) Rv Geol et Sci conn 16:236 (1936)

35a California, an oil empire. Oil Gas J 34, no 25: 61-68 (1935)

Oil Weekly

33 Engineers and geologists find much of interest in results of recent California earthquake. Oil Weekly 69 no 11:33-35 (1933)

Oldham, C.

31 Placer gold discovery in Greenhorn Gulch. M J Ariz 15, no 13:3-4 (1931)

Oldright, G. L.

35 Gold dredging in California. M Metal 16:264 (1935)

Oldroyd, Ida S.

31 (and **Grant**, U. S.) A Pleistocene molluscan fauna from near Goleta, Santa Barbara County, California. Nautilus 44: 91-94 (1931) (*abst*) N Jb 1934, Ref 3:283 (1934) (*abst*) G Zentralbl 46:425 (1932) (*abst*) Pale Zentralbl 1:44 (1932)

Osborn, H. F.

33 *Serbelodon burnhami*, a new shovel-tusker from California. Am Mus Novitates 639:5, 2 figs (1933) (*abst*) N Jb 1934, Ref 3:120 (1935)

Pabst, Adolf

31 The garnets in the glaucophane schists of California. Am Mineralogist 16:327-333 (1931) Min Absts 5:72 (1932) Chem Absts 255:5644 (1931)

31a "Pressure shadows" and measurement of the orientation of minerals in rocks. Am Mineralogist 16:55-70, 14 figs

Pabst, Adolph—Cont.

(1931) (*abst*) Eng Index 1931:893 (1931) (*abst*) N Jb 1931, Ref 2:197-198 (1931) Min Absts 5:74 (1932)

34 Measurement of flow-structures. Am Mineralogist 19:137-143, 8 figs (1934)

36 Vesuvianite from Georgetown, California. Am Mineralogist 21:1-10 (1936) (*abst*) Rv Geol et Sci conn 16:601 (1936)

Pack, R. W.

20 The Sunset-Midway oil field, California. Part I, Geology and oil resources. (*abst*) N Jb 1927, Ref 2, Abt B:114 (1927)

Padover, S. K.

35 Placer mining in California. Pacif Hist Rv 4, no 4:386-392 (1935) *Translation* of Californische Skizzen: (1) Die hydraulischen Goldwaschen by Reyer, E. Deutsche Rundschau (Berlin, 1886, vol XLIX:371-379 (1886)

Palache, Charles

32 (and **Foshag**, W. F.) The chemical nature of Joaquinite. Am Mineralogist 17:308-312, 1 fig (1932) Min Absts 5:191 (1932)

34 Contributions to crystallography; claudetite; minasragrite; samsonite; native selenium; indium. Am Mineralogist 19:194-205 (1934)

Panzer, W.

33 Die kalifornische Sierra Nevada als rumpftreppe. G Rundschau 23a:201-205 (1933) (*abst*) Zs Geom 8:87-88 (1933)

Pardee, J. T.

33 Placer deposits of the western United States. *In* Ore deposits of the western states. Am I M Eng, Lindgren Volume:419-450 (1933) (*abst*) N Jb 1934, Ref 2:712 (1934)

33a (and **Hewett**, D. F.) Manganese in western hydrothermal ore deposits. *In* Ore deposits of the western states. Am I M Eng, Lindgren Volume:671-782, California 671, 672, 680 (1933)

34 Beach placers of the Oregon coast. U S G S circular 8:4-41 (1934)

Parker, Francis L.

31 (and **Cushman**, J. A.) Miocene foraminifera from the Temblor of the east side of the San Joaquin Valley, California. Cushman Lab Foram Res, Contr 7:1-16 (1931) (*abst*) Pale Zentralbl 2:354 (1933) Biol Absts 7:1216, entry 12025 (1933)

Partridge, E. P.

32 (and **Davis**, A. E.) Magnesium and its compounds. U S B M, Min Yb 1932-

Partridge, E. P.—Cont.

1933:777-786, California 779 (1933)... 1934: 1047-1056, California 1048, 1050 (1934)... 1935:1165-1176, California 1166, 1168 (1935)

Parson, B. E.

31 Petroleum developments in California during 1930. *In* Petroleum Dev and Tech, Am I M Eng, Tr 1931:472-486 (1931)

Patmon, C. G.

32 Methods and costs of dredging auriferous gravels at Lancha Plana, Amador County, California. U S B M, Inf Circ 6659:16 pp, 3 figs (1932) (*abst*) U S B M, List Pub July 1, 1932-1933, Suppl:13 (1933)

Patton, J. W.

34 Gems in California. Rocks & Min 9:116-117 (1934)

36 The Mint Canyon agate beds in California. Rocks and Min 11, no 9:156-159 (1936)

Peacock, Martin Alfred

31 The Modoc lava field, northern California. Geog Rv 21:259-275, 15 figs (1931) (*abst*) G Zentralbl 45:294, 488 (1931) Zs Vulkan 15:150 (1933)

Pehrson, Elmer W.

29 Lead. U S B M, Min Res, Calendar Year 1929:231-264, California 236, 240, 241, 242 (1932)... 1930:479-514, California 489, 490, 492 (1933)... 1931:351-375, California 355, 359 (1934)... 1934:79-99, California 86, 87 (1934)... 1935: 75-98, California 85 (1935)

29a Zinc. U S B M, Min Res, Calendar Year 1929: 675-727, California 687 (1932)... Min Yb 1932-1933:67-86, California 73 (1933)... 1934:101-122, California 108 (1934)... 1935:99-122, California 107 (1935) ... 1936:157-179, California 163 (1936)

30 Lead and zinc pigments and salts. U S B M, Min Res, Calendar Year 1930: 113-131, California 125 (1931)... 1931:211-225, California 220, 221 (1932)

33 (and **Meyer**, H. M.) Lead and zinc pigments and zinc salts. U S B M, Min Yb 1932-1933:87-99, California 95 (1933)... 1934:123-137, California 132, 133 (1934)... 1935:123-135, California 132 (1935)

34 Silver yield from copper ores and the effects of 64.64 cent silver on the value of copper ores produced in the United States. U S B M, Inf Circ 6773:15 pp, California 10, 12 (1934)

34a (and **Meyer**, H M) Lead. U S B M, Min Yb 1934:79-99, California 86-87 (1934) ... 1935:75-98, California 85 (1935)

35 (**Furness**, J. W.; and **Meyer**, H. M.) Copper. U S B M, Min Yb 1935:45-73, California 51, 53, 54, 55, 56 (1935)

Pehrson, Elmer W.—Cont.

35a (and **Meyer,** H. M.) Lead and zinc pigments and zinc salts. U S B M, Min Yb 1935:123-135, California 132 (1935)

Pemberton, H., Jr.

91 Analysis of a chromite. Chem News 63:241-242 (1891)

Pemberton, J. R.

35 (and **Allen,** R. E.) How much oil can California produce? Petroleum World 32, no 11:94 (1935)

Penfield, Samuel Lewis

00 On the chemical composition of sulphohalite. Am J Sc (4) 9:425-428 (1900)

Perry, Eugene S.

33 Ventura County investigation. Cal Dp Pub Works, Water Resources, B 46, 46A (1933) (ann) An Bib Ec G 8:162 (1935)

Petar, Alice V.

31 Beryllium and beryl. Cal Dp Nat Res, Div Mines, Mining in California, St Mineralogist's Rp 27:83-97⁻ (1931)

31a (and **Tyler,** P. M.) Molybdenum. U S B M, Min Res, Calendar Year 1931: 75-80, California 77 (1934)

32 (and **Tyler,** P. M.) Rare metals. U S B M, Min Res, Calendar Year 1929: 79-116, California tungsten 98 (1932)

34 (and **Tyler,** P M.) Arsenic. U S B M, Ec P 17:1-35, California 16 (1934)

Petroleum Times

30 California's latest oil field (Venice). Petroleum Times 23:126, 303, map (1930)

30a The Elwood oilfield. Petroleum Times 24:116 (1930)

30b Oilfield of Kettleman Hills. Petroleum Times 24:981, illus (1930)

31 California oil discoveries, important. Petroleum Times 25:742 (1931)

31a California petroleum developments in 1930. Petroleum Times 25:634 (1931)

32 Conservation measures in California. Petroleum Times 27:92, 120, 289 (1932)

32a California's successful unit oilfield development. Petroleum Times 25:427 (1932)

32b Elk Hills oilfield of California. Petroleum Times 28:668 (1932)

32c Californian deep drilling. Petroleum Times 28:130 (1932)

34 California's oil industry in 1933. Petroleum Times 31:201 (1934)

Petroleum World

31 Possibilities of the Sespe at Elwood. Petroleum World 1931 (November):30-31 (1931)

Petroleum World—Cont.

32 Oil possibilities of the Rumsey Hills. Petroleum World 1932 (October):24-28, 45 (1932)

32a Standard ready to test Santa Rosa Island. Petroleum World 1932 (May):15-17, 37 (1932)

33 Gas discovery in northern California. Petroleum World 1933 (February):13 (1933)

33a Depth record broken at Kettleman. Petroleum World 1933 (May):18-22 (1933)

33b Hogan brings a dead oil district to life. Petroleum World 1933 (June):20-22 (1933)

34 Oil well depth record falls again. Petroleum World 1934 (July):25-26, 38 (1934)

34a California crude oil production. Petroleum World, An Rv 1934:51-57 (1934)... 1935:51-56 (1935)

34b California production record, 1933. Petroleum World, An Rv 1934:71-94 (1934)... 1934-1935:71-95 (1935)

34c Natural gasoline plants in California. Petroleum World, An Rv 1934:125, 128 (1934)... 1935:240 (1935)

34d California oil well completions with daily output and dry hole record. Petroleum World, An Rv 1934:58-59 (1934)... 1935:58-59 (1935)

34e California curtailment procedure. Petroleum World, An Rv 1934:157-161 (1934)

34f Asphalt production, 1933. Petroleum World, An Rv 1934:155-157 (1934)

34g California oil field data. Petroleum World, An Rv 1934:96-103 (1934)... 1935: 96-103 (1935)

35 Factors in Del Rey's early collapse. Petroleum World 1935 (August):13-14 (1935)

35a Drilling shows big gain in California. Petroleum World, An Rv 1935:163-172, 177-184 (1935)

36 California oil well completions with daily output and dry hole record. Petroleum World An Rv 1936:72-73 (1936)

36a Oil field developments of the year. Petroleum World An Rv 1936:181-182, 186-190, 194, 200, 204-208 (1936)

36b Three areas in Kern County look grand. Petroleum World An Rv 1936 (December):20-22 (1936)

36c California drilling operations, 1935. Petroleum World An Rv 1936:74 (1936)

36d California crude oil production. Petroleum World, An Rv 1936:65-70, 86-113 (1936)

36e California oil field data. Petroleum World An Rv 1936:114-121 (1936)

Petroleum World (London)

30 Fewer wells in California. Petroleum World (London) 27:157 (1930)

Petroleum World (London)—Cont.

30a The world's deepest hole. Petroleum World (London) 27:218 (1930)

30b Standard Oil of California and its prospects. Petroleum World (London) 27:306 (1930)

30c Items from many fields. United States. Petroleum World (London) 27:410-413, California 411-412 (1930)

30d California: prospects for 1930. Petroleum World (London) 27:62-63 (1930)

30e United States oil industry in 1929. Stocks: California output. Petroleum World (London) 27:146 (1930)

31 Items from many fields. Facts about Kettleman. Petroleum World (London) 28:224 (1931)

32 Menace to control in California. Petroleum World (London) 29:152-153 (1932)

32a The Kettleman unit plan. Petroleum World (London) 29:96 (1932)

33 California and fuel oil. Petroleum World (London) 30:334 (1933)

33a Items from many fields. California oil trade. Petroleum World (London) 30:120 (1933)

33b Items from many fields. Gas in northern California. Petroleum World (London) 30:92 (1933)

36 Known oil reserves. Petroleum World (London) 33:57-62, California 60 (1936)

Phillips, E. R.

32 Sand and gravel. U S B M, Min Res, Calendar Year 1929:359-371, California 359, 362, 365, 366 (1932)... 1930:375-386, California 375, 377, 378, 380, 381 (1932)... 1931:243-250, California 243, 245, 246 (1933)

32a Silica. U S B M, Min Res, Calendar Year 1929:35-40, California 36, 37 (1932)... 1930:111-116, California 112, 113, 114 (1932)... 1931:159-164, California 161, 162 (1933)

Phleger, Fred B. Jr.

33 Notes on certain Ordovician faunas of the Inyo Mountains, California. S Cal Ac Sc, B 32:1-21 (1933)

36 An Ordovician auluroid from California. S Cal Ac Sc, B 35:82-84, pl (1936)

Pierce, A. L.

34 Gold in California. M Mag 24:23-25 (1934) (*abst*) Eng Index 1934:536 (1934)

Pierce, G. G.

32 (Crown, W. J.; and **Howard,** P. J.) Developments in the Long Beach oil field. Cal Dp Nat Res, Div Oil and Gas, California Oil Fields 18 (2):5-26 (1932) (*ann*) An Bib Ec G 7:100 (1935) (*abst*) Rv Geol et Sci conn 14:625 (1934)

Pilsbry, H. A.

34 Pliocene fresh water fossils of the Kettleman Hills and neighboring California oil fields. Nautilus 48:15-17 (1934) Biol Absts 9:929, entry 8322 (1935)

35 Mollusks of the fresh-water Pliocene beds of the Kettleman Hills and neighboring oil fields, California. Nat Ac Sc, Phila Pr 86:541-570, 6 pls (1935) Biol Absts 9:1895, entry 17181 (1935)

Piper, Arthur Maine

32 Investigations of underground-water problems in Arizona, California, New Mexico, and Oregon. Am Geop Union, 13th Annual Meeting, Tr 308-310, Nat Res Council (1932)

32a (Gale, H. S.; and **Thomas,** H. E.) Geology of the Mokelumne River Basin, California. U S G S typescript Rp 377 pp (1932) (*abst*) Nat Res Council, B 98:192-193 (1935)

33 Fluctuations of water surface in observation wells and at stream gaging stations in the Mokelumne area, California, during the earthquake of December 20, 1932. Am Geop Union, Tr 14:471-475 (1933)

34 (Pritchett, H. C.; and **Bue,** C. D.) Seepage loss and gain of the Mokelumne River, California. U S G S Memorandum for the Press (Mimeographed) Ph 85246 (1934)

Plein, L. N.

36 (and **Clark,** J. B.) Fuel briquets. U S B M, Min Yb 1936:649-661, Californit 653, 657 (1936)

Porter, William W. II

32 Lower Pliocene in Santa Maria district, California. Am As Petroleum G, B 16:135-143 (1932) (*ann*) An Bib Ec G 5:140 (1932) (*abst*) Rv Geol et Sci conn 12:270 (1932) (*abst*) N Jb 1933, Ref 3:1203-1204 (1933) (*abst*) G Zentralbl, Abt A 47:397 (1932) (*abst*) Inst Pet Tech, J 18:197A (1932) (*abst*) Eng Index 1932:952 (1932)

33 Influence of speed of migration of oil in water encroachments at Casmalia, California. Am As Petroleum G, B 17:1133-1136 (1933) (*ann*) An Bib Ec G 6:294 (1934) (*abst*) G Zentralbl 51:253 (1934)

Posnjak, E.

33 (and **Bramlette,** M. N) Zeolitic alteration of pyroclastics. Am Mineralogist 18:167-171, 1 fig (1933) Min Absts 5:357 (1933)

Potbury, Susan S.

33 A Pleistocene flora from San Bruno, San Mateo County, California. Carnegie Inst, Wash, Pub 415:25-44, 4 pls, 2 figs

Potbury, Susan S.—Cont.

(1932) (*abst*) N Jb 1934, Ref 3:359-360 (1934) Biol Absts 9: 435, entry 3730 (1935) (*abst*) Pale Zentralbl 4:314 (1934)

35 La Porte flora of Plumas County. Carnegie Inst, Wash, Pub 465:29-81, 19 pls (1935) (*abst*) Pan Am G 63:378-379 (1935) (*abst*) G Soc Am, Pr 1935:416 (1936)

Powell, E. Baden

34 Geology of Santa Barbara Mesa field. Petroleum World 1934 (November): 22-30, 44, map (1934)

36 Geology of the El Segundo oil field. Petroleum World 1936 (February):15-17 (1936)

Powers, Howard A.

32 The lavas of the Modoc Lava Bed quadrangle, California. Am Mineralogist 17:253-294, 1 fig, 6 pls (1932) (*abst*) Rv Geol et Sci conn 13:82-83 (1933) (*Review*) Sc Progress 29:121 (1934) (*abst*) N Jb 1932 Ref 2, Band 2:878 (1932) Chem Absts 26:5042 (1932)

Powers, Sidney

31 Occurrence of petroleum in North America. Am I M Eng, Tr 1931:489-533, California 523-527 (1931) Petroleum World (London) 28:53-59, California 53-57 (1931)

Pressler, Edward D.

29 The San Fernando group in the Las Posas-South Mountain district, Ventura County, California. (*abst*) N Jb 1931, Ref 3:228-229 (1931)

Preston, H. M.

31 Report on Fruitvale oil field. Cal Dp Nat Res, Div Oil and Gas, California Oil Fields 16, no 4:5-24, 6 pls (1931) (*abst*) Rv Geol et Sci conn 13:234 (1933) (*ann*) An Bib Ec G 8:154 (1935)

32 North Belridge oil field. Cal Dp Nat Res, Div Oil and Gas, California Oil Fields 18, no 1:5-24 (1932) (*abst*) Inst Pet Tech, J 20:45A (1934) (*abst*) Eng Index 1934:754 (1934) (*abst*) Rv Geol et Sci conn 15:209-210 (1935)

Pritchett, H. C.

34 (**Bue,** C. D. and **Piper,** A. M.) Seepage loss and gain of the Mokelumne River, California. U S G S Memorandum for the Press (mimeographed) (Ph85246) (1934)

Putnam, William C.

31 (**Davis,** W. M.; and **Richards,** G. L.) Elevated shore lines of the Santa Monica Mountains, California. (*abst*) G Soc Am, B 42:309-310 (1931)

Putnam, William C.—Cont.

33 (and **Livingston,** A.) Geological journeys in southern California. Los Angeles Jr. Coll, Pub 1:104 pp, 56 figs, 1 pl (1933)

35 (and **Grant,** U. S.) Barrancos and arroyos in California. (*abst*) Pan Am G 63:307-308 (1935)

Quayle, Ernest H.

31 Corals of the genus *Caryophyllia* from California. (*abst*) G Soc Am, B 42: 369 (1931) (*abst*) Pan Am G 54:239 (1930)

32 Fossil corals of the genus *Turbinolia* from the Eocene of California. San Diego Soc N H 7:91-109, 12 figs, 1 pl (1932) (*abst*) Rv Geol et Sci conn 12:495 (1932) (*abst*) Pale Zentralbl 2:103 (1932) Biol Absts 8:1587, entry 14543 (1934)

33 Statistical study of some California species of Cardita. (*abst*) Pan Am G 59: 375 (1933) (*abst*) G Soc Am, Pr 1933 1:389 (1934) (*abst*) N Jb 1935, Ref 3:257 (1935)

34 (and **Grant,** U. S.) A new Middle Miocene Neptunea from California. Nautilus 47:91-93 (1934)

Raguin, E.

34 Les failles vivantes en Californie. Terre Vie (11):603-611, 7 figs (1934)

Rambo, A. I.

33 (and **Kennard,** T. G.) Occurrence of rubidium, gallium, and thallium in lepidolite from Pala, California. Am Mineralogist 18:454-455 (1933) (*ann*) An Bib Ec G 6:194 (1934)

Rand, William W.

31 Preliminary report of the geology of Santa Cruz Island, Santa Barbara County, California. Cal Dp Nat Res, Div Mines, Mining in California, St Mineralogist's Rp 27:214-219 (1931) (*abst*) Rv Geol et Sci conn 13:289 (1933) (*abst*) G Zentralbl 46:244 (1932) (*abst*) Eng Index 1931:655 (1931)

Randolph, Gladys C.

34 Turquoise trails. Oregon Mineralogist 2, no 2:3-4, 20-21 (1934)

35 Santa Catalina Island. The Mineralogist 3, no 8:7-8 (1935)

Rankin, Wilbur D.

30 A method of subsurface correlation in the Los Angeles basin. Stanford Univ, Micropaleontology B 2:3-5 (1930)

31 Some notes on the foraminifera of Newport Lagoon, Orange County, California. Stanford Univ, Micropaleontology B 2:75-76 (1931)

Ransome, Frederick Leslie

32 Geologic considerations affecting the choice of a route for the Colorado River aqueduct. (*abst*) G Soc Am, B 43:233 (1932) (*abst*) Pan Am G 55:369 (1931)

Rathbun, Mary Jane

32 A new species of *Cancer* from the Pliocene of the Los Angeles basin. Wash Ac Sc, J 22:19, 1 fig (1932) (*abst*) Rv Geol et Sci conn 12:241 (1932) (*abst*) Pale Zentralbl 3:117 (1933)

32a Fossil pinnotherids from the California Miocene. Wash Ac Sc, J 22:411-413, 11 figs (1932) (*abst*) Rv Geol et Sci conn 13:254 (1933) (*abst*) Pale Zentralbl 4:156 (1934)... 4:367 (1934) Biol Absts 7: 1477, entry 14666 (1933)

Raymond, Louis C.

35 Small native sulphur deposits associated with gossans. M Metal 16:414 (1935)

Read, Charles B.

33 Fossil floras of Yellowstone National Park, Part I. Coniferous woods of Lamar River flora. Carnegie Inst, Wash, Pub 416:21-68, 16 pls (1933)

Redfield, A. H.

32 Asphalt and related bitumens. U S B M, Min Res, Calendar Year 1929:523-567, California 525, 536, 537, 542-543, 548, 558, 559, 565 (1932)... 1930:205-246, California 206, 209, 211, 213, 215, 216, 219, 220, 221, 225, 235, 236, 239, 240, 241, 242, 244 (1932)... 1931:215-235, California 219, 231, 234-235 (1933)... 1932-1933:555-564, California 557 (1933)... 1934:765-774, California 767 (1934)... 1935:871-882, California 873, 874 (1935)... Min Yb 1936:775-788, California 775, 777, 778, 779 (1936)

Reed, Ralph D.

25 (and **Rogers,** A. F.) Sand-calcite crystals from Monterey County, California. (*abst*) N Jb 1925, Ref 2, Abt A:281 (1925)

30 Recent sands of California. J G 38: 223-245, 15 figs (1930) (*abst*) Rv Geol et Sci conn 11:78 (1930) (*abst*) Eng Index 1930:823-824 (1930)

31 Calcareous beds of San Pedro Hills (*abst*) G Soc Am, B 42:310-311 (1931)

31a A siliceous shale formation from southern California. (*abst*) N Jb 1931, Ref 3:571-572 (1931)

31b Microscopic subsurface work in oil fields of the United States. Am As Petroleum G, B 15:731-754, California 741-746 (1931) (*abst*) Inst Pet Tech, J 17: 424A (1931) (*abst*) Zs Prak G 40:94 (1932)

Reed, Ralph D.—Cont.

32 Section from the Repetto Hills to the Long Beach oil field. Int G Cong, Guide Book 15:30-34 (1932)

33 Geology of California. Am As Petroleum G:355 pp, 60 figs (*Review* by Olaf P. **Jenkins**) Ec G 28:697-700 (1933), *Review* Inst Pet Tech, J 19:815 (1933) (*abst*) Petroleum Times 30:21 (1933) (*abst*) G Mag 71:92-93 (1934) (*abst*) Pale Zentralbl 4:268 (1934) (*abst*) Eng Index 1933:535 (1933)

33a Oil-bearing Pliocene beds of southern California. (*abst*) Pan Am G 59:231 (1933)

33b Pleistocene history in the Carpinteria district. (*abst*) Pan Am G 59:306 (1933) (*abst*) G Soc Am, Pr 1933:304 (1934)

33c Santa Margarita conglomerate of Temblor Range. (*abst*) Pan Am G 59: 312 (1933) (*abst*) G Soc Am, Pr 1933 1: 309-310 (1934)

34 Unsolved geological problems of the Pacific Coast. Pan Am G 61:179-187 (1934)

34a Miocene orogenies in California Coast Ranges. (*abst*) Pan Am G 62:303 (1934)... 63:303-304 (1935) (*abst*) G Soc Am, Pr 1935:328 (1936)

35 (and **Hollister,** J. S.) Paleogeology of southern California. (*abst*) Am As Petroleum G, B 19:1841 (1935) (*ann*) An Bib Ec G 8:358 (1936)

35a Geology of California: some corrections. Am As Petroleum G, B 19: 1819-1824 (1935)

35b (and **Hollister,** J. S.) Geology of the Transverse Ranges. (*abst*) Am As Petroleum G, B 19:136 (1935)

35c Miocene breccias of Santa Barbara district. (*abst*) Pan Am G 64:76-77 (1935) (*abst*) G Soc Am, Pr 1935:353 (1936)

35d Tertic limestones of San Rafael Mountains. (*abst*) Pan Am G 64:76 (1935) (*abst*) G Soc Am, Pr 1935:352 (1936)

36 (**Hollister,** J. S.; **Ashauer,** H.) Sedimentation und faltung im sudlichen Kalifornien. Stille-Festschrift 1936:232-233 (1936)

36a (and **Hollister,** J. S.) Structural evolution of southern California. Am As Petroleum G, B 20:1529-1692, 57 figs, 9 pls, 6 tables, map (1936)

Reeds, Chester Albert

33 The Long Beach, California, earthquake. Nat Hist 33:340-341 (1933)

Reeside, John B. Jr.

32 The upper Cretaceous ammonite genus *Barroisiceras* in the United States. U S G S, P P 170:9-29, pls 3-10 (1932)

32a Stratigraphic nomenclature in the United States. Int G Cong, Guide Book 29:7 pp, 10 pls (1932)

Reiche, Parry

34 Geology of Lucia quadrangle, Monterey County. (*abst*) G Soc Am, Pr 1934: 313 (1935) (*abst*) Pan Am G 61:310 (1934)

Reid, H. F.

33 Conception of focus. Nat Res Council, B 90:104-105 (1933)
33a The mechanics of earthquakes. The elastic rebound theory. Regional strain. Nat Res Council, B 90:87-103 (1933)

Reinhart, Phillip W.

33 Tertiary Arcidae of Pacific slope of North America. (*abst*) Pan Am G 59: 373-374 (1933) (*abst*) G Soc Am, Pr 1933: 388 (1934) Stanford Univ Abstracts Dissertations 8:102-105 (1932-1933)
35 Classification of the pelecypod family Arcidae. Musee royal d'Histoire naturelle de Belgique, B Tome XI, no 13: 1-68, California 8, 17, 25, 27, 42, 53 (1935)
35a (and **Schenck,** H. G.) Oligocene arcid pelecypods of genus *Anadara.* (*abst*) Pan Am G 63:373-374 (1935)

Requa, Lawrence K.

32 Description of the property and operations of the Lewiston Dredge, Lewiston, California. U S B M, Inf Circ 6660:1-15 (1932)

Resser, Charles F.

30 Cambrian fossils from the Mojave Desert, California. Smiths Misc Col 81, no 2:1-14, 3 pls (1930)

Reusch, Ethel Grace

33 (and **Reusch,** H. E.) Historic spots in California. (*Review*) Geog J 81:455 (1933)

Reusch, H. E.

33 (and **Reusch,** E. G.) Historic spots in California. (*Review*) Geog J 81:455 (1933)

Reyer, Eduard

86 Californische Skizzen. (1) Die hydraulischen Goldwaschen. Deutche Rundschau (Berlin, 1886, vol XLIX:371-379 (1886) (*Translation*) Padover, S. K. Placer mining in California. Pacif Hist Rv 4, no 4:386-392 (1935)

Rich, John Lyon

34 Problems of the origin, migration, and accumulation of oil. *In* Problems of petroleum geology. Am As Petroleum G, Sidney Powers Memorial Volume:337-345 (1934)

Richard, L. M.

22 California clays require special treatment to meet metallurgical demands. Pacific M News I, No. 1:13 (1922)

Richards, G. L.

31 (**Davis,** W. M.; and **Putnam,** W. C.) Elevated shore lines of the Santa Monica Mountains, California. (*abst*) G Soc Am, B 42:309-310 (1931)
35 Revision of some California species of *Astrodapsis*. San Diego Soc N H, Tr 8:59-66, 1 pl (1935) (*abst*) Rv Geol et Sci conn 15:426-427 (1935)
35a *Astrodapsis* faunal zones of California upper Miocene and lower Pliocene formations. (*abst*) Pan Am G 63:374-375 (1935) (*abst*) G Soc Am, Pr 1935:412 (1936) (*abst*) Pan Am G 63:374 (1935)

Richter, Charles F.

31 Earthquake of January 28, 1931 (California). Seism Soc Am, B 21:284 (1931)
31a (with **Wood,** H. O.) Recent earthquakes near Whittier, California. Seism Soc Am, B 21:183-203 (1931) (*ann*) An Bib Ec G 5:191 (1932) (*abst*) G Zentralbl, Abt A 48:496 (1933)
31b (and **Wood,** H. O.) A study of blasting records in southern California. Seism Soc Am, B 21:28-46 (1931) (*abst*) N Jb 1931, Ref 2:661 (1931) (*abst*) G Zentralbl, Abt A 48:477 (1933)
31c (and **Gutenberg,** B.) On supposed discontinuities in the mantle of the earth. Seism Soc Am, B 21:216-223 (1931)
32 (**Gutenberg,** B.; and **Wood,** H. O.) The earthquake in Santa Monica Bay, California, on August 30, 1932. Seism Soc Am, B 22:138-154, 1 fig, 2 pls (1932) (*abst*) N Jb 1933, Ref 2:573-574 (1933)
33 (and **Wood,** H. O.) A second study of blasting recorded in southern California, Seism Soc Am, B 23:95-110, 2 pls (1933) (*abst*) Rv Geol et Sci conn 13:620 (1933) (*abst*) N Jb 1934, Ref 2:550 (1934) (*abst*) G Zentralbl, Abt A 47:177 (1932)

Ricketts, A. H.

31 Manner of locating and holding mineral claims in California (with forms). Cal Dp Nat Res, Div Mines, B 106:20 pp (1931) (*ann*) An Bib Ec G 5:1 (1932)

Ridgway, Robert H.

31 Pyrites. U S B M, Inc Circ 6523: 1-27, California 10 (1931)
32 Sulphur and pyrites. U S B M, Min Res, Calendar Year 1929:175-194, California 180, 189-190 (1930)... 1930:117-135, California 119, 122, 130, 131 (1931)... 1931: 131-158, California 146, 154 (1932)

Ridgway, Robert H.—Cont.

32a Manganese and manganiferous ores. U S B M, Min Res, Calendar Year 1930:297-332, California 305 (1932)... 1931: 153-184, California 160 (1932)... Min Yb 1932-1933:243-258, California 244 (1933)... 1934:399-416, California 401 (1934)... 1935: 465-484, California 470 (1935)... 1936:425-441, California 436, 437 (1936)

33 Manganese, general information. U S B M, Inf Circ 6729:30 pp (1933)

33a (and **Mitchell,** A. W.) Sulphur and pyrites. U S B M, Min Yb 1932-1933:669-786, California 671, 677, 681 (1933)... 1934: 907-928, California 918-919, 924 (1934)... 1935:1011-1028, California 1020, 1024 (1935)

35 Chromite. U S B M, Min Yb 1935: 521-533, California 523 (1935)

36 Manganese and manganiferous ores. U S B M, Min Yb 1936:425-441, California 436, 437 (1936)

36a (and **Mitchell,** A. W.) Pyrites. U S B M, Min Yb 1936:909 (1936)

36b (and **Umhau,** J. B.) Tungsten. U S B M, Min Yb 1936:447-455, California 449 (1936)

Rieber, Frank

31 Results of elastic-wave surveys in California and elsewhere. (*abst*) Rv Geol et Sci conn 12:472 (1932) (*abst*) G Zentralbl 44:226 (1931)

Ritter, George Joseph

34 (and **Mitchell,** R. L.) Composition of three fossil woods mined from the Miocene auriferous gravels of California. Am Chem Soc, J 56:1603-1605, 1 fig July (1934) Chem Absts 28:5376 (1934)

Roalfe, G. A.

32 Quarrying and crushing methods and costs at the Santa Catalina Island Quarry of Graham Bros., Inc., Santa Catalina Island, California. U S B M, Inf Circ 6609:15 pp, 9 figs (1932) (*abst*) U S B M, List Pub July 1, 1932 to June 30, 1933. Suppl:12 (1933)

Robertson, G. R.

30 Natural gasoline—a new miracle. Chemicals 34:27-28 (1930)

Robinson, Bruce H.

32 (and **Marsh,** H. N) Advantages of flowing wells through tubing. *In* Petroleum Dev and Tech, Am I M Eng, Tr 1932:301-305 (1932)

Robinson, T. W.

31 (with **Stearns,** H. T.; and **Taylor,** G. H.) Geology and water resources of the Mokelumne area, California. (*ann*) An Bib Ec G 4:333 (1930) (*abst*) G Zentralbl 43:375-376 (1930-1931)

Robinson, W. W., Jr.

35 Outlook for California natural gasoline. Petroleum World An Rv 1935:237-239, 243, 248, 252, 257 (1935)

Robotham, C. A.

34 Mining limestone by a caving method at Crestmore mine of the Riverside Cement Co., Crestmore, California. U S B M, Inf Circ 6795:1020 (1933) (*abst*) Eng Index 1934:640 (1934)

Rogers, Austin Flint

25 (and **Reed,** R. D.) Sand-calcite crystals from Monterey County, California. (*abst*) N Jb 1925, Ref 2, Abt A:281 (1925)

25a Euhedral magnesite crystals from San Jose, California. (*abst*) N Jb Ref Band 2, Abt A 1925:284-285 (1925)

31 Chromite in the dunite of northwestern Siskiyou County, California. (*abst*) Pan Am G 55:368-369 (1931) (*abst*) G Soc Am, B 43:232 (1932) (*ann*) An Bib Ec G 5:287 (1932)

31a Castanite, a basic ferric sulphate from Knoxville, California. Am Mineralogist 16:396-404 (1931) (*ann*) An Bib Ec G 4:217 (1931) (*abst*) Am Mineralogist 16:115 (1931) Min Absts 5:96 (1932) Chem Absts 25:5645 (1931)

31b Granite pegmatite from Salt Creek, Tulare County, California. (*abst*) Am Mineralogist 16:116 (1931)

31c Geological history of Lone Hill, Santa Clara County, California. (*abst*) G Soc Am, B 42:316 (1931)

31d Periclase from Crestmore, near Riverside, California, with a list of minerals from this locality. (*abst*) N Jb 1931, Ref 1:80-81 (1931)

32 Anauxite as a secondary mineral in some volcanic rocks of California and Arizona. (*abst*) Pan Am G 58:72-73 (1932) (*abst*) G Soc Am, B 44:159-160 (1933) (*ann*) An Bib Ec G 6:244 (1934)

32a Euhedral gold crystals from Mariposa County, California. (*abst*) Am Mineralogist 17:1115 (1932)

32b Sanbornite, a new barium silicate mineral from Mariposa Co., California. Am Mineralogist 17:161-172 (1932) (*ann*) An Bib Ec G 5:35 (1932) (*abst*) Rv Geol et Sci conn 12:568-569 (1932) Min Absts 5:145 (1932) (*abst*) N Jb 1932, Band 1, Ref 1:408 (1932) Chem Absts 26:4772 (1932)

32c Sanbornite, a new barium disilicate mineral from Mariposa County, California. Cal Dp Nat Res, Div Mines, Mining in California, St Mineralogist's Rp 28:84 (1932) (*abst*) Am Mineralogist's Rp 17:117 (1932)

33 Cleavage and parting in quartz. (*abst*) Am Mineralogist 18:111-112 (1933)

34 Salton volcanic domes of Imperial County. (*abst*) Pan Am G 61:372-373

Rogers, Austin Flint—Cont.

(1934) (*abst*) G Soc Am, Pr 1934:328 (1935)

34a Unique occurrence of vein quartz in Mariposa County, California. (*abst*) Pan Am G 61:370-371 (1934) (*abst*) G Soc Am, Pr 1934:327-328 (1935) (*ann*) An Bib Ec G 8:57 (1935)

Rogers, H. O.

30 (Tryon, F. G.; and Mann, L.) Coal. U S B M, Min Res, Calendar Year 1930: 599-773, California 614, 707 (1932)

34 (and Metcalf, R. W.) Feldspar. U S B M, Min Yb 1934:999-1007, California 1005, 1006, 1007 (1934)... 1935:1107-1114, California 1112 (1935)

36 (and Galiher, C.) Feldspar. U S B M, Min Yb 1936:735-743, California 740, 741, 742 (1936)

Romanowitz, C. M.

34 (and Young, G. J.) Gold dredging. Eng M J 135:486-490 (1934)

34a $34 gold stimulates the dredge designer's ingenuity. Eng M J 135:338-341 (1934)

Roscoe, H. E.

76 On two new vanadium minerals. R Soc London, Pr 25:109-112 (1876)

Ross, Clyde P.

33 Quicksilver deposits. *In* Ore deposits of the western states. Am I Eng, Lindgren Volume:652-658, California 654 (1933)

Ross, C. S.

32 (and Kerr, P. F.) Manganese minerals of a vein near Bald Knob, North Carolina. Am Mineralogist 17:1-18, 2 figs, California 13 (1932) Min Absts 5:50-51 (1932)

Ross, Roland Case

35 A new genus and species of pigmy goose from the McKittrick Pleistocene. San Diego Soc N H, Tr 8:107-114, figs (1935) (*abst*) Rv Geol et Sci conn 15: 427-428 (1935)

Roundy, P. V.

32 (Woodring, W. P.; and Farnsworth, H. R.) Geology and oil resources of the Elk Hills, California (including Naval Petroleum Reserve No. 1), U S G S, B 835:82 pp, 8 figs, 22 pls (1932) (*abst*) Rv Geol et Sci conn 13:367, 513 (1933) (*abst*) Eng Index 1933:818 (1933)

Rousch, G. A.

32 Platinum group metals. Min Ind 41:407-418, California 407 (1933)... 42:456-469, California 459 (1934)... 43:460-473, California 463 (1935)

Rousselot, N. A.

33 (and Silent, R.) The killing and well history of Milham Elliott No. 1 and Continental Elliott 12-8 wells on the North Dome of Kettleman Hills, California. In Petroleum Dev and Tech, Am I M Eng, Tr 1933:91-111 (1933)

Russell, Richard Dana

31 (and Vander Hoof, V. L.) A vertebrate fauna from a new Pliocene formation in northern California. Cal Univ, Dp G, B 20:11-21, 7 figs (1931) (*abst*) G Zentralbl 46:194 (1932) Pale Zentralbl 1:137 (1932) Biol Absts 8:1035, entry 9537 (1934)

32 Nomlaki tuff, its origin and mode of emplacement. (*abst*) G Soc Am, B 43:186-187 (1932) (*abst*) Pan Am G 57:234 (1932)

33 Fossil pearls from the Chico formation of Shasta County, California. (*abst*) N Jb 1933, Ref 3:423-424 (1933) (*abst*) Pale Zentralbl 5:115-116 (1934)

Russell, Richard Joel

30 Basin range structure and stratigraphy of the Warner Range, northeastern California. (*abst*) Peterman's Mitt 76:274 (1930)

31 Do fault patterns indicate type of displacement? (*abst*) G Soc Am, B 42:300 (1931)

31a Dry climates of the United States. Cal Univ, Pub Geog 5:1-41 (1931)... 245-274 (1932)

32 Land forms of the San Gorgonio Pass, southern California. Cal Univ Pub Geog 6:23-121, 42 figs, 1 map (1932)

32a Landslide lakes of the northwestern Great Basin. (*abst*) Zs Geom 7:268 (1932)

33 Alpine land forms of western United States. G Soc Am, B 44:927-950 (1933)

Russell, S.

31 The romance of Signal Hill. Oil B 17:588-596 (1931)

Rutten, Von L.

35 Alte Land-und Meeresverbindungen in West-Indien und Zentralamerika. G Rundschau, Bd XXVI, Heft 1/2:65-94, 2 pls, 3 figs in text (1935)

Salvatori, Henry

33 Correlation of reflection seismograph records in California. Am As Petroleum G, B 17:257-267 (1933) (*abst*) Rv Geol et Sci conn 13:497 (1933) (*ann*) An Bib Ec G 6:135 (1935) (*abst*) Eng Index 1933:540 (1933)

Sampson, Reid J.

31 (with Tucker, W. B.) Feldspar, silica, andalusite, and cyanite deposits of

Sampson, Reid J.—Cont.

California. Cal Dp Nat Res, Div Mines, Mining in California, St Mineralogist's Rp 27:407-458 (1931) (*ann*) An Bib Ec G 4:264 (1931)

31a (and **Tucker**, W. B.) Los Angeles Field Division; San Bernardino County. Cal Dp Nat Res, Div Mines, Mining in California, St Mineralogist's Rp 26:203-334, 25 figs, 3 pls (1930) (*abst*) Eng Index 1931:898 (1931)

32 Mineral resources of a part of the Panamint Range. Cal Dp Nat Res, Div Mines, Mining in California, St Mineralogist's Rp 28:357-376 (1932) (*ann*) An Bib Ec G 6:9 (1934)

32a Economic mineral deposits of the San Jacinto quadrangle. Cal Dp Nat Res, Div Mines, Mining in California, St Mineralogist's Rp 28:3-11 (1932) (*ann*) An Bib Ec G 5:12 (1932)

32b Placers of southern California. Cal Dp Nat Res, Div Mines, Mining in California. 28:245-255 (1932) (*ann*) An Bib Ec G 5:54 (1932) (*abst*) Eng Index 1932: 632 (1932)

32c (and **Tucker**, W. B.) Los Angeles field division, Ventura County. Cal Dp Nat Res, Div Mines, Mining in California, St Mineralogist's Rp 28:247-277 (1932) (*abst*) Eng Index 1933:722 (1933)

33 (and **Tucker**, W. D.) Gold resources of Kern County. Cal Dp Nat Res, Div Mines, M G, J, St Mineralogist's Rp 29: 231-239 (1934) (*ann*) An Bib Ec G 7:41 (1935) (*abst*) Eng Index 1934:533 (1934)

35 Mineral resources of a portion of the Perris Block, Riverside County, California. Cal Dp Nat Res, Div Mines, M G, J 31:507-521 (1935)

Sanborn, Frank

32 Prospecting for vein deposits. Cal Dp Nat Res, Div Mines, Mining in California, St Mineralogist's Rp 28:214-218 (1932) (*ann*) An Bib Ec G 5:54 (1932)

San Diego State College, Delvers Quarterly

36 Preliminary report of the geology of the southern half of the Cuyamaca State Park and adjacent districts, San Diego County, California. San Diego State Coll, Delvers Quarterly 1, no 4:28 (mimeographed) illus (1936)

Santmyers, R. M.

31 Boron and its compounds. U S B M, Inf Circ 6499:37 pp. 1 fig (1931) (*abst*) U S B M, List Pub, 1910-1932:150 (1933)

31a Quartz and silica. Part I. General summary. U S B M, List Pub 1910-1932: 149 (1933)... Inf Circ 6472:15 pp, 1 fig (1931)

Santmyers, R. M.—Cont.

31b Quartz and silica. Part II. Quartz, quartzite and sandstone. U S B M, Inf Circ 6473:20 pp (1931) (*abst*) U S B M, List Pub 1910-1932:149 (1933)

31c Quartz and silica. Part III. Sand and miscellaneous silicas. U S B M, Inf Circ 6474:17 pp (1931) (*abst*) U S B M, List Pub 1910-1932:149 (1933)

31d Umber, sienna, and other brown earth pigments. U S B M, Inf Circ 6504: 14 pp (1931) (*abst*) U S B M, List Pub 1910-1932:150 (1933)

31e Lead. Min Ind 39:356-384, California 366 (1931)... 40:321-350, California 330-331 (1932)

32 Gypsum. U S B M, Min Yb 1932-1933:617-628, California 621, 623, 625 (1933)

32a (and **Stoddard**, B. H.) Barite and barium products. U S B M, Min Res, Calendar Year 1929:209-218, California 209, 211 (1932)... 1930:291-301, California 291, 292, 293, 294, 301 (1932)... 1931:289-296, California 290, 291 (1933)... Min Yb 1932-1933:753-761, California 754, 756, 757 (1933)... 1935:1125-1136, California 1129 (1935)

32b (and **Middleton**, J.) Gypsum. U S B M, Min Res, Calendar Year 1929:105-118, California 106 (1932)... 1930:87-100, California 90, 91, 92, 93, 94, 95 (1932)... 1931:191-203, California 193, 195, 199 (1933)

Sappar, Karl Theodor

16 Bertrage zur geographie der tatigen vulkane. Zs Vulkan 3:65-197, California 83 (1916-1917)

Sauer, Carl

32 Land forms in the Peninsular Range of California as developed about Warner's Hot Springes and Mesa Grande. (*Review*) Geog Rv 20:529-530 (1930) (*abst*) Petermann's Mitt 78:106 (1932) (*abst*) Zs Geom Band VII, Heft 4/5:246-248 (1932)

Savage, J. L.

31 (with **Berkey**, C. P.; **Louderback**, G. D.; **Hinderlider**, M. C.; and **Williams**, I. A.) Report of consulting board on safety of the proposed Pine Canyon dam, Los Angeles County. Cal Dp Pub Works, Div Water Resources May 1931:22 pp, 8 pls, 1 map (1931)

Sawdon, W. A.

36 Handling high-pressure wells in California. Petroleum Eng 6, no 6:51-59, illus (1936)

Sawyer, Edmund O.

35 Geophysical survey by seismic reflection. Petroleum World An Rv 1935: 197-208 (1935) (*abst*) Rv Geol et Sci conn 16:303 (1936)

Schaller, W. T.

05 Notes on some California minerals. (*abst*) N Jb 1905, Band 1:204-205 (1905)

31 Borate minerals from the Kramer district, Mojave Desert, California. (*abst*) N Jb 1931; Ref 1:150-155 (1931) (*abst*) G Zentralbl 44:346 (1931)

31a Hydroboracite from California. Min Absts 4:93-94 (1931)

32 (**Fairchild,** J. G.) Bavenite, a beryllium mineral, pseudomorphous after beryl, from California. Am Mineralogist 17: 409-422 (1932) (*ann*) An Bib Ec G 5:255 (1932) (*abst*) Am Mineralogist 17:114 (1932) Min Absts 5:230-231 (1933) (*abst*) N Jb 1933, Ref 1:235-237 (1933) (by Schaller) (*abst*) Rv Geol et Sci conn 14:97 (1934) Chem Absts 26:5515 (1932)

32a Chemical composition of cuprotungstite. Am Mineralogist 17:234-237 (1932)

33 Pegmatites. *In* Ore deposits of the western states, Lindgren Volume:144-151 (1933) (*abst*) N Jb 1934, Ref 2:206-207 (1934)

35 Monticellite from San Bernardino County, California, and the Monticellite series. Am Mineralogist 20:815-827, 12 tables (1935) (*ann*) An Bib Ec G 8:267 (1936) (*abst*) Rv Geol et Sci conn 16:200 (1936)

Schenck, Hubert Gregory

30 An additional occurrence of *Amphistegina californica.* Stanford Univ, Micropaleontology B 2:43-44 (1930)

31 Cephalopods of the genus *Aturia* from western North America. Cal Univ, Dp G Sc, B 19:435-490, 5 figs, 13 pls (1931) (*abst*) G Soc Am, B 42:369 (1931) (*abst*) Pan Am G 54:238 (1930) (*abst*) Pale Zentralbl 1:118 (1932)

31a (and **Nelson,** R. N.) Eocene algae and stellate orbitoids from the Santa Ynez Range, California. (*abst*) G Soc Am, B 42:371 (1931)

31b (**Kerr,** P. E.; and **Muller** S.) Geology of the Ventura quadrangle, California. (*abst*) G Soc Am, B 42:186-187 (1931) (*abst*) Pan Am G 55:64 (1931)

31c Miocene brown shales of the Kettleman Hills wells, California. (*abst*) G Soc Am, B 42:300 (1931)

31d Diatomaceous shales interbedded with arkose. Stanford Univ Micropaleontology B 2:105-106 (1931)

32 (and **Nomland,** J. O.) Cretaceous beds at Slate's Hot Springs, California. Cal Univ Dp G Sc, B 21:37-49, 4 figs (1932) (*abst*) Rv Geol et Sci conn 12:592 (1932) (*abst*) G Zentralbl, Abt A 48:379-398 (1932) (*abst*) Eng Index 1932:612 (1932)

33 New records of *Discocyclina* in the California Eocene. Stanford Univ, Micropaleontology B 4:72-79 (1933) (*abst*) Pale Zentralbl 15:97-98 (1934)

Schenck, Hubert Gregory—Cont.

33a Significance of *lepidocyclina* in California. (*abst*) Pan Am G 59:319-320 (1933)... 62:156 (1934) (*abst*) G Soc Am, Pr 1933:302 (1934) (*abst*) G Zentralbl 51: 407 (1934)

33b (and **Taliaferro,** N. L.) *Lepidocyclina* in California. Am J Sc 25:74-80 (1933) (*abst*) Pale Zentralbl 3:412 (1933) Biol Absts 9:1886, entry 17081 (1935)

35 What is the Vaqueros formation of California, and is it Oligocene? Am As Petroleum G, B 19:521-536 (1935) (*abst*) 137 (1935) (*ann*) An Bib Ec G 8:150 (1935) (*abst*) Rv Geol et Sci conn 15:252 (1935) (*abst*) Eng Index 1935:784 (1935) (*abst*) G Zentralbl, Abt A, Bb 55:47 (1935)

35a (and **Keen,** A. M.) West American marine molluscan provinces. (*abst*) Pan Am G 63:376-377 (1935) (*abst*) G Soc Am, Pr 1935:413 (1936)

35b (and **Kleinpell,** R. M.) Foraminifera from Gaviota formation. (*abst*) Pan Am G 46:76 (1935) (*abst*) G Soc Am, Pr 1935: 352 (1936)

35c Valid species of the nuculid pelecypod *Acila.* Musee royal d'Historie naturelle de Belgique, B Tome XI, no 14:1-5 (1935)

35d Oligocene problem. (*abst*) Pan Am G 63:75 (1935)

35e (and **Reinhart,** P. W.) Oligocene arcid pelecypods of genus *Anadara.* (*abst*) Pan Am G, 63:373-374 (1935)

36 (and **Kleinpell,** R. M.) Refugian stage of Pacific Coast Tertiary. Am As Petroleum G, B 20:215-225 (1936) (*abst*) B Zentralbl, Abt A, Band 57:50 (1936)

36a Nuculid bivalves of the genus *Acila.* G Soc Am, Sp P no 4:48-50, 82-86, 94-99, 106-110, 112-114, 117-118 (1936) (*abst*) Pan Am G 56:68 (1931) (*abst*) N Jb 1933, Ref 3:1054-1055 (1933) (*abst*) Rv Geol et Sci conn 16:587-589 (1936)

Schenk, Edward Theodore

33 Marine Triassic of central Oregon. (*abst*) Pan Am G 59:374 (1933)

35 A new ammonite genus from the upper Triassic of central Oregon. The American Midland Naturalist 16, no 3: 400-405, California 404 (1935)

Schmeltz, Fred W.

32 Lapis-lazuli in California. Rocks & Min 7:69 (1932)

Schmidt, Ludwig

35 (**Thorne,** H. M.; and **Wilhelm,** C. J.) Disposing of oil field brines. Petroleum World (London) 32:103-106, California 105 (1935)

Schmitt, H. A.

34 (**Locke, A.**; and **Billingsley, P.**) Some ideas on the occurrence of ore in the western United States. Ec G 29:560-576 (1934)

Scholtz, Hermann

31 Die bedeutung makroskopischer gefugentersuchungen fur die rekonstruction fossiler vulkan. Zs Vulcan 14:97-117, pls 15-16, 10 figs (1931)

Schrader, F. C.

33 Epithermal antimony deposits. *In* Ore deposits of the western states. Am I M Eng, Lindgren Volume:658-665, California 661-662 (1933)

Schroter, G. A.

35 A geologist visits the Mojave mining district. Eng M J 136:185-188 (1935) (*ann*) An Bib Ec G 8:55 (1935) (*abst*) Eng Index 1935:520 (1935) (*abst*) G Zentralbl, Abt A, Bd 57:155-156, 269-270 (1936)

Schuchert, Charles

19 Petroliferous provinces. Am I M Eng, B 155:3058-3070 (1919)
33 The Vaqueros formation, lower Miocene of California. Part I. Paleontology by Wayne Loel and W. H. Corey. (*Review*) An J Sc (5) 26:528-529 (1933)

Schuette, Curt Nicolaus

31 Occurrence of quicksilver ore bodies. Am I M Eng, Tech Pub 335:88 pp, 16 figs (1930) (with discussion) Tr (general volume) 1931:403-488, 16 figs (1931)
31a Quicksilver. U S B M, B 335:168 pp, 56 figs (1931) (*abst*) U S B M List of Pub 1910-1932:25 (1933)
31b Quicksilver mining in the United States, Compressed Air 36:3457-3461 (1931)
33 Lahontan quicksilver. Eng M J 134:329-332 (1933)

Schultz, John R.

36 *Plesippus Francescana* (Frick) from the Coso Mountains. (*abst*) G Soc Am, Pr 1935: 419 (1936) (*abst*) Pan Am G 64:79 (1936)

Scofield, C. S.

31 (and **Gale, H. S.**) McKenzie Taylor's genesis of petroleum and coal as applied to Fruitvale field, California. Am As Petroleum G, B 15:709-712 (1931)

Sedelmyer, H. A.

31 Preparation of a new relief map of California. Cal Dp Nat Res, Div Mines, Mining in California, St Mineralogist's Rp 27:73-77, 2 figs, map (1931)

Seeley, E. M.

31 (and **Hopkins, G. R.**) Natural gasoline plants in the United States, January 1, 1932. U S B M, Inf Circ 6635:28 pp (1931)
32 (and **Hopkins, G. R.**) Natural gasoline. U S B M, Stat App, Min Yb 1932-1933:55-65, California 56, 57, 58, 60, 61, 62, 63 (1934)
32a (and **Hopkins, G. R.**) Natural gasoline. U S B M, Min Res, Calendar Year 1929:299-318, California 303, 304, 305, 306, 308, 309, 312, 313, 314, 317, 318 (1932)... 1930:433-456, California 435, 436, 437, 438, 439, 440, 443, 444, 445, 446, 447, 449, 450, 451, 453, 455, 456 (1932)... 1931:331-347, California 331, 334, 335, 336, 338, 344 (1933)

Segerstrom, C. H.

35 Gold mining in California. M Cong J 21, no 4:14, 35 (1935)

Seismological Society of America

31 Seismological notes (California earthquakes). Seism Soc Am, B 21:61, 62, 65, 66, 67, 68, 70, 71, 72, 73, 75, 76, 79, 176, 179, 229, 230, 231, 232, 233, 234, 235, 236, 289, 291, 292, 293, 294, 295, 296, 297, 298, 299, 300, 301, 302, 303, 304, 305, 306, 307, 308 (1931)... 22:172, 175, 251, 252, 256, 288, 289, 290, 291, 292, 295, 296 (1932) ... 23:23, 24, 25, 26, 27, 28, 29, 30, 31, 32, 35, 83, 85, 86, 128, 129, 130, 131, 132, 133, 134, 135, 136, 137, 174, 175, 176, 177 (1933) ... 24:73, 74, 75, 76, 77, 118, 119, 120, 121, 122, 123, 124, 125, 126, 127, 128, 129, 130, 131, 132, 133, 134, 135, 136, 137, 138, 140, 141, 326, 327, 328, 329, 330, 331, 332, 333, 420, 421, 422, 423, 424, 425, 426, 427 (1934) ... 25:96-112, 184-189, 275-279, 387-390 (1935)

Seward, A. C.

32 The redwoods of California: the past and the present. Nature 130:723-726 (1932)

Shannon, Earl V.

30 (and **Short, M. N.**) Violarite and other rare nickel sulphides. (*abst*) Rv Geol et Sci conn 11:158 (1930)

Sharp, R. P.

35 Geology of Ravenna quadrangle. (*abst*) Pan Am G 63:314 (1935) (*abst*) G Soc Am, Pr 1935:336 (1936)

Shaw, E.

31 New gypsum products plant in California. Rock Products 34, no 19:60-64, 4 figs (1931) (*abst*) Eng Index 1931:685 (1931)

Shea, G. B.

32 Natural gasoline. U S B M, Min Yb 1932-1933:535-544, California 541, 543 (1933)... 1934:737-745, California 738, 739, 740, 741, 742, 743, 744 (1934)... 1935:821-831, California 823, 827, 829 (1935)

Shedd, Solon

31 Bibliography of the geology and mineral resources of California to the end of 1929. Cal Dp Nat Res, Div Mines, Geologic Branch, B 104:205 pp (1931) (new edition, revised) 376 pp (1933) (*ann*) An Bib Ec G 4:1 (1931) 6: 170 (1934) (*Review*) Sierra Club B 19:107 (1934)

Shenon, Philip John

33 Copper deposits in the Squaw Creek and Silver Peak districts and at the Alameda mine, southwestern Oregon, with many notes on the Pennell & Farmer and Banfield Prospects. U S G S, Circ 2:34 pp, 5 figs, 6 pls (incl geol maps) (1933)

Shepard, Francis P.

33 Investigation of California submarine canyons. (*abst*) G Soc Am, Pr 1933: 107-108 (1934)

34 (and **McDonald,** G. A.) Sediments of Santa Monica Bay. (*abst*) Pan Am G 61: 317 (1934)

35 Exact configuration of two California submarine canyons. (*abst*) G Soc Am, Pr 1934:106-107 (1935)

35a Continued explorations of California submarine canyons. Am Geop Union, Tr 1935:221-223 (1935)

36 New discoveries from the California submarine canyons. (*abst*) G Soc Am, Pr 1935:104 (1936)

36a Northward continuation of the San Andreas fault. (*abst*) G Soc Am, Pr 1935:105 (1936)

36b The underlying causes of submarine canyons. Nat Ac Sc 22:496-502 (1936)

Shimer, Hervey W.

34 Correlation chart of geologic formations of North America. G Soc Am, B 45:909-935 (1934)

Shore, F. M.

36 Peat. U S B M, Min Yb 1936:663-665, California 665 (1936)

Short, Allan M.

33 A chemical and optical study of piedmontite from Shadow Lake, Madera County, California. Am Mineralogist 18: 493-500 (1933) (*abst*) N Jb. 1934, Ref 1: 449-450 (1934) Min Absts 5:519 (1934) (*abst*) Eng Index 1933:831 (1933)

Short, M. N.

30 (and **Shannon,** E. V.) Violarite and other rare nickel sulphides. (*abst*) Rv Geol et Sci conn 11:158 (1930)

Shrock, R. R.

35 (and **Hunsicker,** A. A.) A study of some great basin lake sediments of California, Nevada and Oregon. J Sed Petrology 5:9-30 (1935)

Shuey, E. T.

32 (and **Clark,** M. B.) Summary of mineral production. U S B M, Stat App, Min Yb 1932-1933:A1-A40, California A13 (1934)... 1934:A1-A41, California A14 (1935)

34 (and **Clark,** M. B.) Summary of mineral resources of the United States. U S B M, Min Res, Calendar Year 1931:A5-A112, California A86 (1934)

35 (**Bagley,** B. W.; and **Hughes,** H. H.) Cement. U S B M, Min Yb 1935:883-909, California 887, 898, 901, 904 (1935)

Siegfus, Stanley S.

35 (and **Cushman,** J. A.) New species of foraminifera from the Kreyenhagen shale of Fresno County, California. Cushman Lab Foram Res, Contr 11:90-96 (1935)

Siemon, J. H.

33 (**Collins,** L. B.; and **Gilbert,** J. C.) Marine oil drilling in California. Petroleum World (London) 30:327-329 (1933) Petroleum World 1933 (October):15-18 (1933)

Silent, R.

33 (and **Rousselot,** N. A.) The killing and well history of Milham Elliott No. 11 and Continental Elliott 12-8 wells on the North Dome of Kettleman Hills, California. *In* Petroleum Dev and Tech, Am I M Eng, Tr 1933:91-111 (1933)

Simonson, R. R.

35 Piedmontite from Los Angeles County, California. Am Mineralogist 20: 737-738 (1935) (*ann*) An Bib Ec G 8:268 (1936)

Simpich, F.

33 Men and gold. Nat Geog Soc, Mag 43:481-518, California 481, 497, 501, 502 (1933)

Simpson, Edward C.

34 Geology and mineral deposits of the Elizabeth Lake quadrangle, California. Cal Dp Nat Res, Div Mines, M G, J, St Mineralogist's Rp 30:371-415 (1934) (*ann*) An Bib Ec G 8:57 (1936) (*abst*) G Zentralbl, Abt, Bd 57:201-202 (1936)

Simpson, G. G.

33 Glossary and correlation charts of North American Tertiary mammal bearing formations. Am Mus, B 67:79-121 (1933) (abst) N Jb 1934, Ref 3:308 (1934)

Sinclair, William John

06 The exploration of the Potter Creek cave, California. (geology and Quaternary vertebrates). (abst) N Jb 1906, Ref 1:125-127 (1906)

Singewald, Joseph T. Jr.

33 Genetic groups of hypogene deposits and their occurrence in the western United States. In Ore deposits of the western states. Am I M Eng, Lindgren Volume:503:524, California 505-506, 512, 516-518, 519-520, 522-524 (1933)

Skeggs, J. H.

33 Skilful work of engineers saved highway in 200,000 cubic yard slide. Cal Dp Pub Works, Cal Highways, April:22 (1935)

Skidmore, W. A.

85 Gravel channels of ancient rivers. M Sc Press 51:6 (1885)

Smith, Allyn G.

33 (and Hanna, G. D.) Two new species of Monadenia from northern California. Nautilus 46:79-86, 2 pls (1933) Biol Absts 8:243, entry 2121 (1934)

Smith, Hampton

34 Origin of some siliceous Miocene rocks of California. (abst) Pan Am G 61:376-377 (1934) (abst) G Soc Am, Pr 1934:334 (1935)

Smith, James Perrin

06 The comparative stratigraphy of the marine Trias of western America. (abst) N Jb 1906, Band 2:98-100 (1906)

09 The paragenesis of the minerals in the glaucophane-bearing rocks of California. (abst) N Jb 1909, Band 1:71-74 (1909)

31 Upper Triassic marine invertebrate faunas of North America. G Zentralbl 44:440-447 (1931)

32 Lower Triassic ammonoids of North America. U S G S, P P 167:199 pp, 81 pls, 1 fig (1932) (abst) Rv Geol et Sci conn 13:443 (1933) (abst) N Jb 1933, Ref 3:598-610 (1933) Biol Absts 8:804, entry 7233 (1934)

Smith, Lewis A.

32 Chromite. U S B M, Min Res, Calendar Year 1929: 203-229, California 205, 206 (1932)... 1930:243-266, California 244-245 (1933)... 1931:103-116, California 104

Smith, Lewis A.—Cont.

(1934)... Min Yb 1932-1933:299-309, California 300 (1933)

32a Manganese and manganiferous ores. U S B M, Min Res, Calendar Year 1929: 275-332, California 282 (1932)

32b Chromium. U S B M, Inf Circ 6566:1-32, 6 figs, California 18, 19-20 (1932)

Smith, Wayne M.

30 Some foraminifera from the Elwood field, Santa Barbara County, California. Stanford Univ, Micropaleontology B 2:5-8 (1930)

Smith, W. S. T.

33 Marine terraces on Santa Catalina Island. Am J Sc 25:123-136 (1933) (abst) Eng Index 1933:536 (1933)

Smyth, H. L.

05 Origin and classification of placers. Eng M J 79:1045-1046, 1179-1180, 1228-1230 (1905)

Snedden, Loring B.

32 Notes on the stratigraphy and the micropaleontology of the Miocene formation in Los Sauces Creek, Ventura County, California. Stanford Univ, Micropaleontology B 3:41-46 (1932)

Snider, L. C.

33 A comparison of old and new oil fields. In Petroleum Dev and Tech; Am I M Eng, Tr 1933:71-86 (1933)

34 Current ideas regarding source beds for petroleum. In Problems of petroleum geology. Am As Petroleum G, Sidney Powers Memorial Volume:51-66 (1934)

36 (and Brooks, B. T.) Probable petroleum shortage in the United States, and methods for its alleviation. Am As Petroleum G, B 20:15-50, California 24, 27-28, 34, 36, 40 (1936)

Somers, George B.

30 Anomalies of vertical intensity compared with regional geology for the State of California. (abst) G Zentralbl 43:421-422 (1930-1931)

Soper, E. K.

32 Limitations of ground water in determining hidden geologic structures. Am As Petroleum G, B 16:335-360 (1932) (abst) Inst Pet Tech, J 18:198A (1932) (abst) Eng Index 1932:952 (1932)

32a (with Grant, U. S.) Geology and paleontology of a portion of Los Angeles, California. G Soc Am, B 42:1041-1067 (1932) (ann) . An Bib Ec G 5:341 (1932) (abst) Pan Am G 57:370-371 (1932) (abst) G Soc Am, B 44:148 (1933) (abst) Pale

Soper, E. K.—Cont.

Zentralbl 4:162 (1934)... 5:91-92 (1934) Biol Absts 9:1857, entry 16826 (1935)

34 (and **Grant,** U. S.) Stratigraphy of western Santa Monica Mountains. (*abst*) Pan Am G 61:308 (1934) (*abst*) G Soc Am, Pr 1934:310-311 (1935)

Soske, J. L.

33 Differences in diurnal variation of vertical magnetic intensity in southern California. Terrestrial Magnetism and Atmospheric Electricity 1933:109-115 (1933) (*abst*) N Jb 1935, Ref 2:485 (1935)

35 (and **Kelley,** V. C.) Wave-built pumice deposits and Salton rhyolitic hills. (*abst*) Pan Am G 63:319-320 (1935) (*abst*) G Soc Am, Pr 1935:341 (1936)

35a Magnetometer survey of southern portion of San Andreas fault. (*abst*) Pan Am G 63:318 (1935) (*abst*) G Soc Am, Pr 1935:340 (1936)

36 (and **Kelley,** V. C.) Origin of the Salton volcanic domes, Salton Sea, California. J G 44:496-509, figs (1936) (*abst*) Rv Geol et Sci conn 16:514-515 (1936) (*abst*) G Zentralbl, Abt A, Bd 58:28 (1936)

Sparks, Neil R.

32 (and **Byerly,** P.) Earthquakes in northern California and the registration of earthquakes at Berkeley, Mount Hamilton, Palo Alto from October 1, 1931 to March 31, 1932. Cal Univ, Seism Sta B 3:53-96 (1933)... April 1, 1932 to September 30, 1932, B 3:97-150 (1933)... October 1, 1932 to March 31, 1933, B 3:151-241 (1935)

33 (and **Byerly,** P.) The first preliminary waves of the California earthquake of June 6, 1932. Seism Soc Am, Eastern Sec, Earthquake Notes 5:254-256, 1 fig (1933) Am Geop Union, Tr 14:254-256 (1933)

33a (and **Byerly,** P.) Earthquakes in northern California and the registration of earthquakes at Berkeley, Mount Hamilton, Palo Alto from October 1, 1931, to March 31, 1932. Cal Univ, Seism Sta B 3:54-96 (1933)

33b (and **Byerly,** P.) Earthquakes in northern California and the registration of earthquakes at Berkeley, Mt. Hamilton, Palo Alto from October 1, 1931, to April 1, 1932 to September 30, 1932. Cal Univ, Seism Sta, B 3:97-150 (1933)... October 1, 1932 to March 31, 1933. Cal Univ, Seism Sta, B 3:151-241 (1935)

35 Building vibrations. Seism Soc Am, B 25:381-387 (1935)

Spicer, H. Cecil

36 Rock temperatures and depths to normal boiling point of water in the

Spicer, H. Cecil—Cont.

United States. Am As Petroleum G, B 20:270-279, California 271, 279 (1936)

Spiers, James

31 Mining methods and costs at Central-Eureka Mine, Amador County, California. U S B M, Inf Circ 6512:12 pp, 18 figs (1931) (*abst*) U S B M, List Pub 1910-1932:151 (1933) (*abst*) Eng Index 1931:674 (1931)

33 Mining methods and costs at Central Eureka mine, California. Explosives Eng 11, no 4:119-123 (1933) U S B M, Inf Circ 6512, 12 pp, 18 figs (1931)

Spilman, C. F.

31 Grass Valley-Nevada City district stages comeback. M J Ariz 15, no 12:3-4, 37 (1931) (*abst*) Eng Index 1931:673-674 (1931)

31a The Hadsel mill and Beebe gold mine. M J Ariz no 15:5-6 (1931)

Spitaler, R.

34 Uber die Erdbeden in Kalifornien. Beitr Geoph 42:321-328 (1934) (*abst*) N Jb 1935, Ref 2:485 (1935)

Staack, J. G.

32 Topographic branch, California. U S G S, An Rp 1932:48 (1932)

Stabler, Herman

32 Conservation branch, California. U S G S, An Rp 1932:62 (1932)

Stalder, Walter

31 New productive horizon in California. Am As Petroleum G, B 15:201 (1931) (*abst*) G Zentralbl 44:478 (1931)

31a Northern California's chance for oil. Petroleum World 28:49-51, 127 (1931)

31b New light on California's early days. Petroleum World 28:32-34, 40 (1931)

32 Commercial oil and gas production in northern California not far off. Petroleum World 29, no 7:13-16, 60 (1932)

32a Structural and commercial oil and gas possibilities of central valley region, California. Am As Petroleum G, B 16: 361-371 (1932) (*ann*) An Bib Ec G 5:145 (1932) (*abst*) Rv Geol et Sci conn 12:618 (1932) (*abst*) N Jb 1933, Ref 3:1204-1205 (1933) (*abst*) G Zentralbl, Abt A 47:413 (1932) (*abst*) Inst Pet Tech, J 18:197A (1932)

33 Gas on Marysville Buttes, Sutter County, California. Am As Petroleum G 17:443 (1933) (*abst*) Eng Index 1933:818 (1933)

33a Northern California development. Petroleum World, An Rv 1933:151-161 (1933)

Standard Oil Bulletin

36 New oil deposit in San Joaquin Valley. Standard Oil, B 24, no 9:1 (1936)

Stanton, T. W.

32 Geologic branch, California. U S G S, An Rp 1932:15-16 (1932)

Stanton, W. Layton, Jr.

31 Geology of the Adelaida quadrangle, California (*abst*) G Soc Am, B 42:301-302 (1931)

32 Structure of portion of southern Santa Lucia range. (*abst*) G Soc Am, B 1344:147-148 (1933) (*abst*) Pan Am G 57: 369-370 (1932)

Stauffer, Clinton R.

31 The Devonian of California. (*abst*) G Zentralbl 44:191 (1931) (*abst*) N Jb 1933, Ref 3:533-534 (1933) Biol Absts 6: 5273 (1932) (*abst*) Pale Zentralbl 4:140-141 (1934)

Stearns, Harold T.

28 Lava beds of National Monument, California. Geog Soc Philadelphia, B 26, no 4:239-253, 12 figs, 1 map (1928)

31 (with **Robinson**, T. W.; and **Taylor**, G. H.) Geology and water resources of the Mokelumne area, California. (*ann*) An Bib Ec G 4:333 (1931) (*abst*) G Zentralbl 43:375-376 (1930-1931)

31a Rainfall and stream run-off in southern California since 1769. Western City 7, no 9:19-20 (1931) (*ann*) An Bib Ec G 4:333 (1931)

Steffa, Don

30 Pumicite, cement's little sister. M J Ariz 14, no 9:7-8 (1930)

32 Gold mining and milling methods and costs at the Vallecito Western Drift Mine, Angels Camp, California. U S B M, Inc Circ 6612:13 pp, 4 figs (1932)

Sterrett, Douglas Bovard

09 Precious stones. (*abst*) N Jb, Ref Band 1, 1909:30-31 (1909)

Stevens, J. B.

30 (and **Jensen**, J.) Water invasion, McKittrick oil field. M Metal 11:470-471 (1930)

31 (and **Jensen**, J.) Water problems of McKittrick oil field. *In* Petroleum Dev and Tech, Am I M Eng, Tr 92:164-167 (1931)

Stewart, J. D.

34 Hydraulic mining again interesting to capital. Eng M J 135:491-493 (1934) (*ann*) An Bib Ec G 7:243 (1934)

Stewart, Katherine C.

30 (and **Stewart**, R. E.) "Lower Pliocene" in eastern end of Puente Hills, San Bernardino County, California. (*abst*) Rv Geol et Sci conn 11:467-468 (1930) (*abst*) Eng Index 1930:823 (1930) Pale Zentralbl 2:222 (1933)

31 (and **Stewart** R. E.) Post-Miocene foraminifera from the Ventura quadrangle, Ventura County. (Twelve new species and varieties from the Pliocene). (*abst*) N Jb 1931, Ref 3:79 (1931) (*abst*) G Zentralbl 43:266-267 (1930-1931) Biol Absts 9:687, entry 6060 (1935)

33 (and **Stewart**, R. E.) Notes on the foraminifera of the type Merced at Seven Mile Beach, San Mateo County, California. San Diego Soc N H, Tr 7:259-272, 2 pls (1933) (*abst*) Pale Zentralbl 4:80-81 (1934) Biol Absts 8:1326, entry 12240 (1934) (*abst*) Rv Geol et Sci conn 15:218 (1935)

Stewart, Ralph B.

31 Gabb's California Cretaceous and Tertiary type lamellibranchs. (*Review*) J G 39:177-178 (1931) (*abst*) N Jb 1933, Ref 3:161-167 (1933) Biol Absts 6:11261 (1932)

Stewart, Roscoe E.

30 (and **Stewart**, K. C.) "Lower Pliocene" in eastern end of Puente Hills, San Bernardino County, California. (*abst*) Rv Geol et Sci conn 11:467-468 (1930) (*abst*) Eng Index 1930:823 (1930) (*abst*) Pale Zentralbl 2:222 (1933)

31 (and **Stewart**, K. C.) Post-Miocene foraminifera from the Ventura quadrangle, Ventura County. (Twelve new species and varieties from the Pliocene.) (*abst*) N Jb 1931, Ref 3:79 (1931) (*abst*) G Zentralbl 43:266-267 (1930-1931) Biol Absts 9:687, entry 6060 (1935)

33 (and **Stewart**, K. C.) Notes on the foraminifera of the type Merced at Seven Mile Beach, San Mateo County, California. San Diego Soc N H, Tr 7:259-272, 2 pls (1933) (*abst*) Pale Zentralbl 4:80-81 (1934) Biol Absts 8:1326, entry 12240 (1934) (*abst*) Rv Geol et Sci conn 15:218 (1935)

Stille, Hans

36 The present tectonic state of the earth. Am As Petroleum G, B 30:849-878 (1936) (*abst*) Rv Geol et Sci conn 16:516 (1936) (*abst*) G Zentralbl, Abt A, Bd 58: 150-151 (1936) (*abst*) Rv Geol et Sci conn 16:516 (1936)

Stipp, T. F.

34 (and **Tolman**, F. B.) Eocene stratigraphy on north side of Simi Valley.

Stipp, T. F.—Cont.

(*abst*) Pan Am G 42:79 (1934) (*abst*) G Soc Am, Pr 1934:393-394 (1935)

Stirton, R. A.

31 (and **Matthew, W. D.**) Osteology and affinities of Borophagus. (*abst*) N Jb 1931, Ref 3:678-679 (1931) (*abst*) G Zentralbl 5:458 (1931) (*abst*) Pale Zentralbl 1:142 (1932)

32 A new genus of *Soricidae* from the Barstow Miocene of California. (*abst*) Pale Zentralbl 1:297 (1932)

33 Critical review of Mint Canyon mammalian fauna and its correlative significance. Am J Sc 26:569-576 (1933) (*abst*) Pan Am G 59:377 (1933) (*abst*) N Jb 1934, Ref 3:476 (1934) (*abst*) G Soc Am, Pr 1933 1:392 (1934)

33a (and **Vander Hoof, V. L.**) *Osteoborus*, a new genus of dogs and its relations to *Borophagus* Cope. Cal Univ, Dp G, B 23:175-182, 3 figs (1933) (*abst*) N Jb 1934, Ref 3:475 (1934)

34 (and **De Chardin, T. P.**) A correlation of some Miocene and Pliocene mammalian assemblages in North America and Asia with a discussion of the Mio-Pliocene boundary. Cal Univ, Dp G, B 23:277-290, 3 tables (1934) (*abst*) N Jb 1935, Ref 3:913 (1935)

35 A review of the Tertiary beavers. Cal Univ, Dp G, B 23:391-458, 142 figs, 1 map, 2 charts (1935)

36 Succession of North American continental Pliocene mammalian faunas. Am J Sc 32 (5):161-206 (1936)

Stock, Chester

30 Oreodonts from the Sespe deposits of South Mountain, Ventura County, California. Carnegie Inst, Wash Pub 404:27-42, 2 figs, 2 pls (1930)

31 Discovery of upper Eocene land mammals on the Pacific Coast. Science 74:577-578 (1931) (*abst*) Pale Zentralbl 3:219 (1933)

32 Is *Felis atrox* of Rancho La Brea, a lion or a tiger? (*abst*) G Soc Am, B 43:290 (1932) (*abst*) Pan Am G 56:70 (1932) (*abst*) Pale Zentralbl 2:390 (1933)

32a An upper Oligocene mammalian fauna from southern California. Nat Ac Sc, Pr 18:550-554 (1932) (*abst*) Pan Am G 58:71 (1932) (*abst*) G Soc Am, B 44:158 (1933) (*abst*) Pale Zentralbl 5:295 (1934)

32b Additions to the mammalian fauna from the Tecuya beds, California. Carnegie Inst, Wash, Contr Paleontology, Pub 418:87-92, 1 pl (1932) (*abst*) Pale Zentralbl 4:52 (1934)... 5:295 (1934) Biol Absts 8:562, entry 5097 (1934)

32c (and **Merriam, J. C.**) The Felidae of Rancho La Brea. Carnegie Inst, Wash, Pub 422:152 figs, 42 pls (1932) (*abst*) N

Stock, Chester—Cont.

Jb 1934, Ref 3:319-320 (1934) Biol Absts 9:491, entry 4313 (1935)

32d Upper Eocene mammals from the Sespe, north of the Simi Valley, California. (*abst*) G Soc Am, B 44:158 (1933) (*abst*) Pan Am G 58:71 (1932) (*abst*) Pale Zentralbl 4:180 (1934)

32e *Hyaenognathus* from Pliocene of Coso Mountains (Inyo County, California) (*abst*) Pan Am G 58:149 (1932) (*abst*) G Soc Am, B, 44:218 (1933) (*abst*) Pale Zentralbl 4:233 (1934)... 5:361 (1934) (*abst*) J Mammalogy 13:263-266, 1 pl (1932)

32f Eocene land mammals on the Pacific Coast. Nat Ac Sc, Pr 18:518-523, 4 figs (1932)

32g Rancho La Brea, a record of Pleistocene life in California. Los Angeles Museum, Pub 1:1-84, 27 figs (1930) (*abst*) Pale Zentralbl 1:175 (1932)

32h Asphalt deposits and Quaternary life of Rancho La Brea. Int G Cong, Guide Book 15:21-23 (1932)

33 Carnivora of the Sespe upper Eocene, Simi Valley region, California. (*abst*) G Soc Am, B 44:199 (1933) (*abst*) Pale Zentralbl 4:232 (1934)

33a Carnivora of the Sespe Oligocene, Las Posas Hills, California. (*abst*) G Soc Am, B 44:199 (1933) (*abst*) Pale Zentralbl 4:232 (1934)

33b Carnivora from the Sespe of the Las Posas Hills, California. Carnegie Inst, Wash, Pub Paleontology 440:29-41, 3 pls (1934)

33c An Amynodont skull from the Sespe deposits of California. Nat Ac Sc, Pr 19:762-767, 1 fig (1933) (*abst*) N Jb 1934, Ref 3:308-309 (1934) Biol Absts 8:2129, entry 19253 (1934)

33d Hyaenodontidae of the upper Eocene of California. Nat Ac Sc, Pr 19:434-440, 12 figs, 1 pl (1933) (*abst*) N Jb 1933, Ref 3:309 (1934) Biol Absts 8:2129, entry 19251 (1934)

33e An Eocene primate from California. Nat Ac Sc, Pr 19:954-959, 1 pl (1933) (*abst*) N Jb 1934, Ref 3:665 (1934) Biol Absts 8:2129, entry 19254 (1934)

33f (and **Hall, E. R.**) The Asiatic genus Eomellivora in the Pliocene of California. J Mammalogy 14:63-65, 1 pl (1933) Biol Absts 8:834, entry 7628 (1934)

33g A miacid from the Sespe upper Eocene, California. Nat Ac Sc, Pr 19:481-486, 1 pl (1933) Biol Absts 8:2129, entry 19252 (1934) (*abst*) N Jb 1933, Ref 3:309 (1934) (*abst*) G Soc Am, Pr 1933:302 (1934) (*abst*) Pan Am G 59:304 (1933) (*abst*) Pale Zentralbl 5:170 (1934)

33h (and **Merriam, J. C.**) Tertiary mammals from the auriferous gravels near Columbia, California. Carnegie

Stock, Chester—Cont.

Inst, Wash, Pub Paleontology 440:1-6, 2 figs (1934)

33i Canid and proboscidean remains from the Ricardo deposits, Mojave desert, California. Biol Absts 7:762, entry 7523 (1933)

33j A peccary from the McKittrick Pleistocene, California. Biol Absts 7:1258, entry 12543 (1933)

33k A tooth of Hipparion mohavense from the Puente formation, California. Biol Absts 7:516, entry 5024 (1933)

34 A Hypertragulid from the Sespe uppermost Eocene, California. Nat Ac Sc, Pr 20:625-629 (1934)

34a Eocene vertebrate faunas from the Sespe north of Simi Valley. (abst) Pan Am G 41:375 (1934) (abst) G Soc Am, Pr 1934:332 (1935)

34b Microsyopsinae and Hyopsodontidae in the Sespe upper Eocene, California. Nat Ac Sc, Pr 20:349-354 (1934)

34c Eocene vertebrate fauna from the Sespe, north of Simi Valley. (abst) Pan Am G 62:79-80 (1934)

34d New genus of rodent from the Sespe. (abst) G Soc Am, Pr 1934:384 (1934)

34e On the occurrence of an Oreodont skeleton in the Sespe of South Mountain, California. Nat Ac Sc, Pr 20:518-523, 2 pls, 2 figs (1934) (abst) Pale Zentralbl 3:440 (1933)

34f A second Eocene primate from California. Nat Ac Sc, Pr 20:150-154 (1934) Biol Absts 9:492, entry 4320 (1935)

34g New Creodonta from the Sespe upper Eocene, California. Nat Ac Sc, Pr 20:423-427 (1934) Biol Absts 9:1919, entry 17496 (1935)

35 Deep-well record of fossil mammal remains in California. Am As Petroleum G, B 19:1064-1068, 2 figs (1935) (abst) Eng Index 1935:501 (1935) (ann) An Bib Ec G 8:358 (1936) (abst) N Jb 1936, Ref 3:160 (1936)

35a Insectivora from the Sespe uppermost Eocene, California. Nat Ac Sc, Pr 21:214-219, 1 pl (1935) (abst) N Jb 1935, Ref 3:907-908 (1935)

35b Exiled elephants of the Channel Islands, California. Sc Mo 41:205-215, 10 figs (1935) (abst) N Jb 1935, Ref 3:1080 (1935)

35c (and Bode, F. D.) Occurrence of lower Oligocene mammal-bearing beds near Death Valley, California. Nat Ac Sc, Pr 21:571-579, 3 pls (1935) (abst) G Zentralbl, Abt A, Bd 57:53-54 (1936)

35d Plesiomiacis, a new creodont from the Sespe upper Eocene, California. Nat Ac Sc, Pr 21:119-123, 1 pl (1935)

Stock, Chester—Cont.

35e Titanothere remains from the Sespe of California. Nat Ac Sc, Pr 21:456-462, 2 pls (1935)

35f New genus of rodent from the Sespe Eocene. G Soc Am, B 46:61-68, 1 pl, 1 fig (1935) Biol Absts 9:1919, entry 17498 (1935)

35g Artiodactyla from the Sespe of the Las Posas Hills, California. Carnegie Inst, Wash, Pub 453:119-125, 1 pl (1935)

35h Sespe faunas. (abst) Pan Am G 63:315 (1935) (abst) G Soc Am, Pr 1935: 337 (1936)

36 Perissodactyla of the Sespe Eocene, California. Nat Ac Sc, Pr 22:260-266, pls (1936)

36a Hesperomeryx, a new artiodactyl from the Sespe Eocene, California. Nat Ac Sc, Pr 22:177-182, pl (1936)

36b Titanotheres from the Titus Canyon formation, California. Nat Ac Sc, Pr 22:656-662, pls (1936)

36c Sespe Eocene didelphids. Nat Ac Sc, Pr 22:122-124, pl (1936)

36d Ice age elephants of the Channel Islands. Westways 28 (June):14-15, illus (1936)

36e When Titans roamed prehistoric Death Valley. Westways 28 (March):28-29, illus (1936)

Stockman, L. P.

30 Submarine geology opens possibilities. (abst) Eng Index 1930:1300 (1930)

31 New oil fields at San Miguelito, California. Oil Gas, J 30, no 22:17 (1931) (ann) An Bib Ec G 4:317 (1931)

35 California needs 200,000,000 barrels of new crude oil each year. Oil Gas J 34, no 25:87-119 (1935)

35a New oil field opened in El Segundo after years of intensive exploration. Oil Gas J 34, no 14:15 (1935)

Stoddard, Blanche H.

32 Mica. U S B M, Min Res, Calendar Year 1929:373-388, California 373, 375 (1932)

32a (and Bowles, O.) Asbestos. U S B M, Min Res, Calendar Year 1929:195-207, California 199 (1932)... 1930:263-275, California 266, 267 (1932)... Min Yb 1934: 1009-1016, California 1012 (1934)

32b (and Santmyers, R. M.) Barite and barium products. U S B M, Min Res, Calendar Year 1929:209-218, California 209, 211, (1932)... 1930:291-301, California 291, 292, 293, 294, 301 (1932)... 1931:289-296, California 290, 291 (1933)... Min Yb 1932-1933:753-761, California 754, 756, 757 (1933)... 1935:1125-1136, California 1129 (1935)

Stoddard, Blanche H.—Cont.

32c (and **Bowles,** O.) Talc and soapstone. U S B M, Min Res, Calendar Year 1930:303-313, California 304, 305, 306, 307, 308 (1932)... 1931:99-110, California 100, 103, 104, 110 (1933)

33 (and **Emery,** A. H.) Talc and ground soapstone. U S B M, Min Yb 1932-1933: 715-722, California 717, 718, 719 (1933)... 1934:975-984, California 976, 977-978 (1934) ... 1935:1069-1081, California 1070, 1072 (1935)... 1936:953-962, California 953, 954, 955 (1936)

Storie, R. E.

36 (and **Weir,** W. W.) A rating of California soils. Cal Univ, Coll Agri, B 599: 158 pp, 67 tab, 4 maps (1936) (*abst*) Rv Geol et Sci 16:136 (1936)

Strong, A. M.

34 (and **Grant,** U. S.) Fossil mollusks from the vertebrate-bearing asphalt deposits at Carpinteria, California. S Cal Sc, B 33:7-11 (1934) (*abst*) Pan Am G 59:375 (1933) (*abst*) G Soc Am Pr 1933 1:390 (1934) (*abst*) Rv Geol et Sci conn 14:244 (1934)

34a (and **Grant,** U. S.) Pliocene and Pleistocene mollusca of Santa Barbara. (*abst*) Pan Am G 42:71-72 (1934) (*abst*) G Soc Am, Pr 1934:386-387 (1935)

Sutherland, J. C.

31 (**Buwalda,** J. P.; and **Gazin,** C. L.) Frazier Mountain; a crysalline overthrust slab without roots, west of Tejon Pass, southern California. (*abst*) G Soc Am, B 42:294-295 (1931) (*abst*) N Jb 1933, Ref 2:251 (1933)

35 Geological investigation of the clays of Riverside and Orange Counties, southern California. Cal Dp Nat Res, Div Mines, M G, J, St Mineralogist's Rp 31: 51-87, maps (1935) (*ann*) An Bib Ec G 8:30, 308 (1936) (*abst*) G Zentralbl, Abt A, Bd 57:343-344 (1936)

Suverkrop, Lew

31 Oil possibilities of Terra Bella district in Tulare County, California. Oil B 17:14-17, 76, 6 figs (1931) (*abst*) Eng Index 1931:1036 (1931)

Swanson, E. B.

32 Natural gas. U S B M, Min Yb 1932-1933:517-533, California 518, 519 (1933)

Swartzlow, Carl R.

35 Ice caves in northern California. J G 43:440-442 (1935)

Symons, Henry H.

28 California mineral production for 1927. Cal Dp Nat Res, Div Mines, B 101:

Symons, Henry H.—Cont.

311 pp (1928)... 1928, B 102:210 pp (1929)... 1929, B 103:231 pp (1930)... 1930, B 105:229 pp (1931) (*abst*) Eng Index 1930:1106 (1930)... 1931:892 (1931)

30 Commercial grinding plants in California. Northern California plant. Cal Dp Nat Res, Div Mines, Mining in California, St Mineralogist's Rp 26:334-340 (1930)

30a California mineral-paint materials. Cal Dp Nat Res, Div Mines, Mining in California, St Mineralogist's Rp 26:148-160 (1930) U S B M, B 370:77-82 (1933)

31 Minerals and statistics. Cal Dp Nat Res, Div Mines, Mining in California, St Mineralogist's Rp 26:347-348, 499-500 (1930-1931)... 27:113-114 (1931)... 30:291-294, 447-450 (1934)... 31:101-102, 221-245 (1935) (*ann*) An Bib Ec G 4:4 (1931)... 27:224-231, 547-549 (1931)... 28:89-90, 226-234, 396-398 (1932)... 29:252-253, 379-381 (1932-1933)... 30:101-102 1934:31:101-102 221-246, 386-391, 527-531 (1935)

31a California mineral production and directory of mineral producers for 1930. Cal Dp Nat Res, Div Mines, B 105:231 pp (1931) (*ann*) An Bib Ec G 5:6 (1932)

32 California mineral production and directory of mineral producers for 1931. Cal Dp Nat Res, Div Mines, B 107:229 pp, ill (1932)... 1932, B 109:200 pp (1933)

32a The pan, rocker, and sluice box. Cal Dp Nat Res, Div Mines, Mining in California, St Mineralogist's Rp 28: 205-213 (1932)

35 Museum, with index of mineral collection in museum. Cal Dp Nat Res, Div Mines, M G, J, St Mineralogist's Rp 31:228-245 (1935)

Taber, Stephen

33 The location of earthquake epicenters. Science n s 78:283 (1933)

Taff, Joseph A.

33 Geology of McKittrick oil field and vicinity, Kern County, California. Am As Petroleum G, B 17:1-15 (1933) (*abst*) Rv Geol et Sci conn 13:511-512 (1933) (*ann*) An Bib Ec G 6:97 (1934) (*abst*) Inst Pet Tech, J 19:161A-162A (1933) (*abst*) Pan Am G 57:313 (1932)

34 Physical properties of petroleum in California. *In* Problems of petroleum geology. Am As Petroleum G, Sidney Powers Memorial Volume:177-234 (1934) (*ann*) An Bib Ec G 7:311 (1934)

35 Geology of Mount Diablo. G Soc Am, B 46:1079-1100, 1 pl, 1 fig (1935) (*abst*) G Zentralbl, Abt A, Bd 57:369 (1936) (*abst*) Rv Geol et Sci conn 16:349-350 (1936)

35a (**Hanna,** G. D.; and **Cross,** C. M.) Chico Cretacic at type locality. (*abst*)

Taff, Joseph A.—Cont.

Pan Am G 64:72 (1935) (*abst*) G Soc Am, Pr 1935:348 (1936)

  36 Discussion of papers on Mt. Diablo. G Soc Am 46:2043-2045 (1936)

Taliaferro, Nicholas Lloyd

  31 Analcite diabase and related rocks in California. (*abst*) G Soc Am, B 42: 296-297 (1931) (*abst*) Pan Am G 54-73 (1930)

  32 (and **Turner,** R. E.) Lithophysae-bearing rhyolites in the southern Santa Lucia Range. (*abst*) Pan Am G 55:374 (1931) (*abst*) G Soc Am, B 43:237 (1932)

  32a Stratigraphy of the bedrock complex of the Sierra Nevada of California. (*abst*) G Soc Am, B 43:233-234 (1932) (*ann*) An Bib Ec G 5:238 (1932) (*abst*) Pan Am G 55:369-370 (1931)

  32b Bedrock complex of the Sierra Nevada, west of the southern end of the Mother-Lode. (*abst*) G Soc Am, B 44: 149-150 (1933) (*abst*) Pan Am G 57:371-372 (1932)

  33 Contraction phenomena in cherts. (*abst*) Pan Am G 59:305-306 (1933) (*abst*) G Soc Am, Pr 1933:303 (1934) (*abst*) Rv Geol et Sci conn 15:358-359 (1935)

  33a The relation of volcanism to diatomaceous and associated siliceous sediments. Cal Univ, Dp G, B 23:1-56 (1933) (*abst*) Eng Index 1933:291 (1933) (*abst*) Zs Vulkan 16:149-150 (1935)

  33b (and **Schenck,** H. G.) *Lepidocyclina* in California. Am J Sc 25:74-80 (1933) (*abst*) Pale Zentralbl 3:412 (1933) Biol Absts 9:1886, entry 17081 (1935)

  35 Geology of San Simeon, Adelaida, and Paso Robles quadrangles. (*abst*) Pan Am G 63:316 (1935) (*abst*) G Soc Am, Pr 1935:338 (1936)

Taylor, C. A.

  30 (with **Blaney,** H. F.; and **Young** A. A.) Rainfall penetration and consumptive use of water in Santa Ana River valley and Coastal Plain, California. Cal Dp Nat Res, Div Water Res, B 33: 158 pp (1930) (*ann*) An Bib Ec G 4:333 (1931)

Taylor, Edward

  34 (and **Woodford,** A. O.) Longitudinal profiles of streams. (*abst*) G Soc Am, Pr 1933 1:307 (1934)

Taylor, Frank Bursley

  31 Correlation of Tertiary mountain ranges in the different continents. G Soc Am, B 41:431-473, 4 pls (1930) (*abst*) N Jb 1931, Ref 2:616-618 (1931)

Taylor, George F.

  33 Scarp-ramp in northern Owens Valley. (*abst*) Pan Am G 59:311-312 (1933) (*abst*) G Soc Am, Pr 1933 1:309 (1934)

Taylor, G. H.

  31 (**Stearns,** H. T.; and **Robinson,** T. W.) Geology and water resources of the Mokelumne area, California. (*ann*) An Bib Ec G 4:333 (1930) (*abst*) G Zentralbl 43:375-376 (1930-1931)

Ten Eyck, Richard G.

  36 Late Tertiary foraminifera of San Jose Hills, Los Angeles County, California. (*abst*) G Soc Am, Pr 1935:364 (1936)

Thalman, H. E.

  35 Bibliography and index to new genera species and varieties of foraminifera for year 1933. J Paleontology 9:715-743 (1935)

Theller, J. H.

  14 Hydraulicking on Klamath River. M Sc Press 108:523-526 (1914)

Thom, Emma M.

  35 Bibliography of North American geology 1933 and 1934. U S G S, B 869: 389 pp (1935)

Thomas, H. E.

  32 (**Gale,** H. S.; and **Piper,** A. M.) Geology of the Mokelumne River basin, California. U S G S typescript Rp 377 pp (1932) (*abst*) Nat Res Council, B 98: 192-193 (1935)

Thompson, A. P.

  30 Finding the lost Vulture Lode. M J Ariz 14, no 13:8-11, 28-30 (1930)

Thompson, David G.

  20 The Mojave desert region, California. (*abst*) Rv Geol et Sci conn 11:426-427 (1930) (*abst*) Peterman's Mitt 77:327 (1931) (*abst*) Zs Geomorphology 5:280-281 (1929)

Thompson, Warren O.

  33 Observations on the stratification of beach deposits. (*abst*) G Soc Am, B 44: 171 (1933)

  35 Original structure of beaches. Stanford Univ, Abstracts Dissertations 10:77-82 (1934-1935)

Thoms, C. C.

  31 Operations in district No. 2, 1930. Cal Dp Nat Res, Div Oil and Gas, California Oil Fields 16, no 3:35-41 (1931)... 1931, 17, no 3:26-31 (1932)... 1932, 18, no 3:25-31 (1933)

Thorne, H. M.

35 (Schmidt, L., and Wilhelm, C. J.) Disposing of oil field brines. Petroleum World (London) 32:103-106, California 105 (1935)

Thorpe, W. H.

32 Petroleum bacteria and the nutrition of Pailopa petrolei. Nature 130:437 (1932)

Tieje, A. J.

33 (and Cassell, D.) Megafauna and Microfauna of the Pleistocene and Pliocene formations of southern California as revealed in a deep well near Ventura. (abst) Pan Am G 59:376 (1933) (abst) G Soc Am, Pr 1933 1:390-391 (1934)

Tolman, C. F.

31 Geology of upper San Francisco Bay region with special reference to a salt water barrier below confluence of Sacramento and San Joaquin Rivers. Cal Dp Pub Work, Div Water Resources, B 28: 309-360 (1932)

33 The foothill copper belt of California. XVI Int G Cong, Wash, Copper resources of the world 1933:247-249, 2 pls (1933)

Tolman, F. B.

34 (and Stipp, T. F.) Eocene stratigraphy on north side of Simi Valley. (abst) Pan Am G 42:79 (1934) (abst) G Soc Am, Pr 1934:393-394 (1935)

Townsend, R. H.

33 Method and cost of quarrying limestone at plant of Calaveras Cement Co., San Andreas, California. U S B M, Inf Circ 6610:11 pp, 5 figs (1933) (abst) U S B M, List Pub, July 1, 1932 to June 30, 1933, Suppl:12 (1933)

Trask, Parker D.

31 (and Wu, C. C.) Analyses of oil and gas from distillation of recent sediments. (ann) An Bib Ec G 3:135 (1931) (abst) N Jb 1931, Ref 2:513 (1931)

31a Sedimentation in the Channel Islands region, California. Ec G 26:24-43, 6 figs (1931) (ann) An Bib Ec G 4:91 (1931) (abst) Rv Geol et Sci conn 12:146 (1932) (abst) N Jb 1931, Ref 2:740-741 (1931)... 1932, Ref 2:89-290 (1932) (abst) G Zentralbl 45:289 (1931) (abst) Eng Index 1931:1036 (1931) (abst) Zs Prak G 39:180 (1931) Chem Absts 25:1773 (1931)

33 (and Hammar, H. E.) Source beds in Mesozoic rocks west of Sacramento River, California. (abst) Pan Am G 59: 229 (1933)

33a (and Hammar, H. E.) Some relations of the organic constituents of sedi-

Trask, Parker D.—Cont.

ments to the formation of petroleum. (abst) Wash Ac Sc, J 23:568 (1933)

34 (and Hammar, H. E.) Preliminary study of source beds in late Mesozoic rocks on west side of Sacramento Valley, California. Am As Petroleum G, B 18:1346-1373 (1934) (abst) Wash Ac Sc, J 24:491-492 (1934) (abst) N Jb 1935, Ref 2:688 (1935) (ann) An Bib Ec G 8:154 (1935) (ann) An Bib Ec G 7:312 (1935) (abst) Rv Geol et Sci conn 15:207-208 (1935)

35 The organic content of some Tertiary formations in California. Am As Petroleum G, B 19:135 (1935)

35a (and Hammar, H. E.) Organic content of sediments. (abst) Wash Ac Sc 25:508 (1935) Am Petroleum Inst, Productions B 214 (1934) Oil Gas J 33, no 27, 28, 29:43-45, 40-41, 36-39 (1934) Petroleum World (London) 32:19-20 (1935)

36 Proportion of organic matter converted into oil in Santa Fe Springs field, California. Am As Petroleum G, B 20: 245-257, table (1936) (abst) Rv Geol et Sci conn 16:404 (1936)

Tryon, F. G.

· 32 (Mann, L.; Young, W. H., and Bennit, H. L.) Coal. U S B M, Min Res, Calendar Year 1930:599-733, California 614, 707 (1932)

32a (and Mann, L.) U S B M, Min Res, Calendar Year 1929:673-858, California 700, 703, 743 (1932)

33 (Young, W. H.; Berquist, F. E.; Mann, L., and Bennit, H. L.) Coal. U S B M, Min Res, Calendar Year 1931: 415-510, California 426, 472 (1933)

33a (with Young, W. H.; and Corse, J. M.) Fuel briquets. U S B M, Min Res, Calendar Year 1931:61-71, California 65, 67 (1933)

34 (Mann, L.; Young, W. H.; and Bennit, H. L.) Coal. U S B M, Min Yb, Stat App, Calendar Year 1932:373-454, California 376, 423 (1934)... 1933:281-360, California 286, 323 (1935)

Tucker, W. B.

31 Notes on mining activity in Inyo and Mono counties in July, 1931. Cal Dp Nat Res, Div Mines, Mining in California, St Mineralogist's Rp 27:543-545 (1931)

31a (with Sampson, R. J.) Feldspar, silica, andalusite, and cyanite deposits of California. Cal Dp Nat Res, Div Mines, Mining in California, St Mineralogist's Rp 27:407-458 (1931) (ann) An Bib Ec G 4:264 (1931)

31b (and Sampson, R. J.) Los Angeles Field Division; San Bernardino

Tucker, W. B.—Cont.

County. Cal Dp Nat Res, Div Mines, Mining in California, St Mineralogist's Rp 26:203-334, 25 figs, 3 pls (1930) (*abst*) Eng Index 1931:898 (1931)

32 (and **Sampson, R. J.**) Los Angeles field division, Ventura County. Cal Dp Nat Res, Div Mines, Mining in California, St Mineralogist's Rp 28:247-277 (1932) (*abst*) Eng Index 1933:722 (1933)

33 (and **Sampson, R. J.**) Gold resources of Kern County. Cal Dp Nat Res, Div Mines, M G, J, St Mineralogist's Rp 29:271-339 (1933) (*abst*) Eng Index 1934:533 (1934) (*ann*) An Bib Ec G 7:41 (1935)

34 South of the Tehachapi gold mining makes new gain. Eng M J 135:517-521 (1934)

34a Current mining activity in southern California. Cal Dp Nat Res, Div Mines, M G, J, St Mineralogist's Rp 30: 310-329 (1934)

35 Mining activity at Soledad Mountain and Middle Buttes, Mojave mining district, Kern County. Cal Dp Nat Res, Div Mines, M G, J 31:465-483 (1935)

35a Mojave mining district. Western Miner & Prospector 2, no 1:3-4 (1935)

36 Gold mining in the Mojave District; California. M Metal 17:82-85 (1936)

Turner, Henry Ward

22 The Wilshire gold mine. Eng M J 114:888-890 (1922)

Turner, R. E.

31 (and **Taliaferro, N. L.**) Lithophysae-bearing rhyolites in the southern Santa Lucia Range. (*abst*) Pan Am G 55:374 (1931) (*abst*) G Soc Am, B 43:287 (1932)

Turrentine, J. W.

31 Potash. Min Ind 40:445-457, California 445, 446 (1932)... 41:419-435, California 420, 421 (1933)... 42:470-487, California 472 (1934)... 43:474-489, California 475 (1935)

Tyler, Paul M.

31 Magnesite. U S B M, Inf Circ 6437:53 pp, California 4, 7-11, 13, 14, 22-26 (1931) (*abst*) U S B M, List Pub 1910-1932:147 (1933)

31a Magnesium compounds (other than magnesite). U S B M, Inf Circ 6406:19 pp, 1 fig (1931) (*abst*) U S B M, List Pub 1910-1932:145 (1933)

31b (and **Myer, H. M.**) Mercury. U S B M, Min Res, Calendar Year 1931:191-209, California 192, 195, 198-201 (1934)

31c (and **Petar, A. V.**) Molybdenum. U S B M, Min Res, Calendar Year 1931: 75-80, California 77 (1934)

Tyler, Paul M.—Cont.

31d Magnesite. Min Ind 39:385-398, California 387, 388 (1931)

32 Abrasive and industrial diamonds. U S B M, Inf Circ 6562:25 pp (1932) (*abst*) U S B M, List Pub 1910-1932: 154 (1933)

32a Magnesium and its compounds. U S B M, Min Res, Calendar Year 1929: 119-138, California 120, 121, 122, 123, 125, 130, 135, 136 (1932)... 1930:181-203, California 183, 184, 185, 187, 195, 200 (1932)... 1931:263-277, California 265, 266, 267, 275 (1933)

32b Mercury. U S B M, Min Res, Calendar Year 1929:117-142, California 120, 121, 122, 123, 126-127 (1932)... 1930:31-56, California 32, 37, 39-42 (1933)

32c (with **Petar, A. V.**) Rare metals. U S B M, Min Res, Calendar Year 1929: 79-116, California tungsten 98 (1932)

34 (and **Metcalf, R. W.**) Clay. U S B M, Min Yb 1934:873-887, California 876 (1934)... 1935:977-993, California 980 (1935)

34a (and **Petar, A. V.**) Arsenic. U S B M, Ec P 17:1-35, California 16 (1934)

35 Minor nonmetals: graphite, greensand, kyanite, mineral wool, monazite, olivine, strontium minerals, and vermiculite. U S B M, Min Yb 1935:1213-1236, California 1213, 1214, 1225, 1228, 1229, 1231, 1232 (1935)... 1936:1057-1073, California 1057, 1063, 1065, 1066, 1073 (1936)

35a (and **Metcalf, R. W.**) Gypsum. U S B M, Min Yb 1935:949-966, California 951, 953 (1935)

35b Sodium sulphate. U S B M, Inf Circ 6833:1-40, California 14 (1935)

Ulrich, Franklin P.

35 A progress report of the California seismological program of the Coast and Geodetic Survey. Seism Soc Am, B 25: 349-361 (1935)

35a. The California strong-motion program of the United States Coast and Geodetic Survey. Seism Soc Am, B 25: 81-96 (1935)

Umhau, J. B.

36 (and **Ridgway, R. H.**) Tungsten. U S B M, Min Yb 1936:447-455, California 449 (1936)

Union Oil Company Bulletin

35 Progress in petroleum geology. Union Oil Company B 1:6-14 (1935)

36 Petroleum geology in California. Union Oil Company B 3:2-5 (1936)

U S Bureau of Mines

31 Mining methods and costs at the Central-Eureka Mine, Amador County, California. U S B M, Inf Circ 6512 (1931)

U. S. Bureau of Mines—Cont.

32 Mineral resources of the United States, 1930. Part II—Nonmetals, 876 pp, Washington, Govt. Print Off (1932)

33 Mineral resources of the United States, 1931. Part I—Metals, Washington, Govt Print Off (1933)

33a Mineral resources of the United States, 1931. Part II—Nonmetals, Washington, Govt Print Off (1933)

U S Department of the Interior, Geological Survey

34 Mineral resources and possible industrial development in the region surrounding Boulder Dam. U S Dp Interior, Geological Survey, Bur Rec 1934:1-27 (1934)

34a Geology and occurrence of petroleum in the United States. Petroleum Investigation, Hearings, H. Res. 441, Part 2, pp 869-1086 (1934)

Uren, Lester C.

35 (Jomercq, J. Jr.; and Mejea, J.) Large diameter wells are indicated. Petroleum World (London) 32:260-262 (1935)

36 Drilling and production progress in California fields. Petroleum World An Rv 1936:39-64, 248, 250 (1936)

Uwatoko, Kunio

32 Genesis of oil by high radial axial pressure. Am As Petroleum G, B 16: 1029-1037, 1 fig (1932) (abst) Rv Geol et Sci conn 13:364 (1933)

Valentine, W. W.

31 (and Cushman, J. A.) Shallow water foraminifera from the Channel Islands of southern California. (abst) N Jb 1931, Ref 3:77 (1931)

35 Semitropic gas field. Am As Petroleum G, B 19:1843-1844 (1935) (abst) An Bib Ec G 8:358 (1936)

Van Amringe, E. V.

32 Mining in Pasadena. Miner Soc S Cal, B 1, no 3:1-2 (1932)

33 The gem minerals of San Diego County, California. Miner Soc S Cal, B 2, no 7:1-4 (1933) Min Absts 5:281 (1933)

34 Bentonite, neptunite, and joaquinite. Oregon Mineralogist 2, no 11:9-10 (1934)

35 Fine colemanite specimens found in California. The Mineralogist 3, no 1:51 (1935)

Van Couvering, Martin

30 So this is Venice! Oil B 16:1151-1157 (1930)

Vanderburg, W. O.

31 Methods and costs of concentrating

Vanderburg, W. O.—Cont.

tungsten ores at Atolia, San Bernardino County, California. U S B M, Inf Circ 6532:12 pp, 4 figs (1931)

31a Mining methods and costs at Argonaut Mines. M Cong J 17:48-55, 10 figs (1931) U S B M, Inf Circ 6311: 14 pp, figs (1930)

35 Tungsten. U S B M, Inf Circ 6821: 1-31, California 11 (1935)

35a Mining and milling tungsten ores. U S B M, Inf Circ 6852:78 pp, 17 figs, California 8-11, 25-31 (1935) (abst) G Zentralbl, Abt A, Bd 58:284 (1936)

VanderEike, Paul

36 Sharktooth Hill. The Pacific Mineralogist 3, no 2:18, 27 (1936)

Vander Hoof, V. L.

31 (and Russell, R. D.) A vertebrate fauna from a new Pliocene formation in northern California. Cal Univ Dp G, B 20:11-21, 7 figs (1931) (abst) Pale Zentralbl 1:137 (1932) Biol Absts 8:1035, entry 9537 (1934)

31a Boraphagus littoralis from the marine Tertiary of California. Cal Univ Dp G, B 21:15-24, 3 pls (1931) (abst) Pale Zentralbl 2:64 (1932) Biol Absts 8:563, entry 5099 (1934)

33 Additions to the fauna of the Tehama upper Pliocene of northern California. Am J Sc (5) 25:382-384 (1933) Biol Absts 9:1857, entry 16829 (1935)

33a Pliocene vertebrate fauna from Sierra foothills of central California. (abst) Pan Am G 59:376-377 (1933) (abst) G Soc Am, Pr 1933 1:391 (1934)

33b (and Stirton, R. A.) Osteoborus, a new genus of dogs, and its relation to Borophagus Cope. Cal Univ, Dp G, B 23: 175-182, 3 figs (1933) (abst) N Jb 1934, Ref 3:475 (1934)

33c A skull of Pliohippus tantalus from the later Tertiary of the Sierra foothills of California. Cal Univ, Dp G, B 23:183-194, 1 pl, 5 figs (1933) (abst) N Jb 1934, Ref '3:316-317 (1934) Pale Zentralbl 5:299 (1934) Biol Absts 9:962, entry 8646 (1935)

34 Pleistocene vertebrates from northern California. (abst) Pan Am G 62:69 (1934) (abst) G Soc Am, Pr 1934:383-384 (1935)

34a Seasonal banding in asphalt deposits. (abst) Pan Am G 61:374-375 (1934)

35 Seasonal banding in an asphalt deposit at McKittrick. (abst) G Soc Am, Pr 1934:332 (1935)

35a Nature and distribution of Desmostylus, a marine Tertic mammal. (abst) Pan Am G 64:80 (1935) (abst) G Soc Am, Pr 1935:420 (1936)

Van Orstrand, C. E.

32 On the correlation of isogeothermal surfaces with the rock strata. Physics 2, no 3:139-153 (1932)

34 Temperature gradients. *In* Problems of petroleum geology. Am As Petroleum G, Sidney Powers Memorial Volume:989-1021 (1934)

34a Some possible applications of geothermics to geology. Am As Petroleum G, B 18:13-38, California 21, 24, 25, 30, 33 (1934)

35 Normal geothermal gradient in the United States. Am As Petroleum G, B 19:78-115, California 80, 96, 100, 111-112 (1935)

Vanossi, R.

35 La industria del Yodo. Anal Soc Cientif Argent 118, 1934 Entrege 2:105-106 (1934) (*abst*) N Jb 1935, Ref 2:575 (1935)

Vaughan, Thomas Wayland

32 Notes on investigation on modern marine sediments in California. Nat Res Council, B 89:74-79 (1932)

32a Rate of sea cliff recession on the property of the Scripps Institution of Oceanography at La Jolla, California. Science n s 75:250 (1932)

Vickery, Frederick P.

31 Pleistocene history of southern Coast Ranges of California. (*abst*) Pan Am G 56:234 (1931)

Vokes, H. E.

33 New species of *Haliotis* from the Pliocene of southern California. (*abst*) G Soc Am, Pr 1933 1:373 (1934) J Paleontology 9:251-252 (1935)

34 Stratigraphic position of *Turritella andersoni* zone, north of Coalinga. (*abst*) Pan Am G 62:78 (1934) (*abst*) G Soc Am, Pr 1934:393 (1935)

35 The genus *Velates* in the Eocene of California. Cal Univ, Dp G, B 23:381-390, 6 tables (1935) (*abst*) N Jb 1935, Ref 3:903 (1935)

35a Notes on the variation and synonomy of *Ostrea idriaensis* Gabb. Cal Univ, Dp G, B 23:291-304, 3 pls (1935)

36 (and **Clark,** B. L.) Summary of marine Eocene sequence of western North America. G Soc Am, B 47:851-878, 2 pls, 3 figs (1936)

36a Middle Eocene molluscan faunas of the Vallecitos and Coalinga area. (*abst*) G Soc Am, Pr 1935:411 (1936) (*abst*) Pan Am G 63:372 (1935)

36b The gastropod fauna of the intertidal zone at Moss Beach, San Mateo County, California. The Nautilus 50:46-51 (1936)

Vonsen, M.

29 Death Valley and the borates of California. Rocks & Min 13:73 (1929) (*abst*) N Jb 1931, Ref 1:88 (1931)

32 (**Irving,** J.; and **Gonyer,** F. A.) Pumpellyite from California. Am Mineralogist 17:338-342 (1932) (*abst*) Rv Geol et Sci conn 13:84 (1933) Min Absts 5:233 (1933)

35 The discovery of borates in California. The Mineralogist 3, no 12:21-23 (1935)

36 (and **Hanna,** G. D.) Borax Lake, California. Cal Dp Nat Res, Div Mines M G, J St Mineralogist's Rp 32:99-108, 5 text figs (1936)

Wailes, C. D., Jr.

33 (and **Horner,** A. C.) Earthquake damage analyzed by Long Beach officials. Eng News-Rec 110:684-686 (1933) (*abst*) Canada, Dom Obs, Pub, Bib Seism 10:334 (1933) (*abst*) Rv Geol et Sci conn 14:175-176 (1933-1934)

Walcott, Charles Doolittle

12 Cambrian brachiopoda. U S G S Monograph 50:Pt 1 Text 11-872 (1912)

Wallace, K. C.

32 Temblor: past, present, future: Purman well stimulates action. Cal Oil World, 24, no 45:7-8 (1932)

Ward, George W.

31 A chemical and optical study of the black tourmalines. Am Mineralogist 16:145-190 (1931)

Warner, Thor

30 What the drill reveals of subsurface geology at Venice. Petroleum World 1930 (August):78-80 (1930)

31 Mercury deposit in Coso range, Inyo County. (*abst*) Eng Index 1930:1076 (1930) Chem Absts 25:2670 (1931)

Wartenweiler, Otto

36 The new mill of the Golden Queen. Eng M J 137:327-335 (1936)

Washburne, Chester W.

32 Premonitory formations. (*abst*) Pan Am G 57:69-70 (1932)

34 (and **Lahee,** F. H.) Oil-field waters (foreword) *In* Problems of petroleum geology. Am As Petroleum G, Sidney Powers Memorial Volume:833-840 (1934)

Waterfall, Louis N.

29 A contribution to the paleontology of the Fernando group, Ventura County, California. Cal Univ, Dp G, B 18:71-92 pls 5-6, 1 fig (1929) (*ann*) An Bib Ec G 2:129 (1929) (*abst*) G Zentralbl 41:105

Waterfall, Louis N.—Cont.

(1930) (*abst*) Eng Index 1929:879 (1930) (*abst*) N Jb 1931, Ref 3:227-228 (1931)

Waters, Aaron Clement

33 Summary of the sedimentary, tectonic, igneous and metalliferous history of Washington and Oregon. *In* Ore deposits of the western states. Am I M Eng, Lindgren Volume:253-265 (1933)

35 (and **Campbell, C. D.**) Mylonites from San Andreas fault zone near Crystal Springs lakes. Am J Sc (5) 29:473-503 (1935) (*abst*) G Soc Am, Pr 1934:325 (1935) (*abst*) Pan Am G 61:319-320 (1934) (*abst*) Rv Geol et Sci conn 15:359-360 (1935) (*abst*) Sc Progress 30:307 (1936) Chem Absts 29:6182 (1935)

Watkins, S. L.

17 El Doradoite. Am Mineralogist 2:26 (1917)

Watts, Arthur S.

31 Feldspar. Min Ind 39:664-667, California 665, 66 (1931)... 40:606-608, California 606, 607 (1932)... 41:570-572, California 570 (1933)... 42:629-632, California 630 (1934)... 43:631-633, California 631, 632 (1935)

Weatherbee, D'Arcy

06 A hydraulic mine in California. M Sc Press 93:296-298 (1906)

Weaver, Charles Edwin

31 Stratigraphic relations of the Domengine and Markely formations in the Antioch, Vacaville, and Napa quadrangles. (*abst*) G Soc Am, B 42:305 (1931)

32 Geologic cross section through the Coast Ranges immediately north of San Francisco Bay. (*abst*) Pan Am G 58:69 (1932) (*abst*) G Soc Am, B 44:155 (1933)

33 Early Pliocene diastrophism in the Coast Ranges of northern California. (*abst*) G Soc Am, Pr 1933:117 (1934)

Weaver, Donald K.

31 Encroachment of edge water at Santa Fe Springs. *In* Petroleum Dev and Tech. Am I M Eng, Tr 1931:157-162 (1931) M Metal 11:472-474 (1930)

Webb, Robert W.

35 Tetradymite from Inyo Mountains, California. Am Mineralogist 20:399-400 (1935) (*abst*) N Jb 1935, Ref 1:445 (1935) (*abst*) An Bib Ec G 1935:28 (1936)

35a Opportunities for meteorite discoveries in the western United States. The Mineralogist 3, no 3:5-6 (1935)

35b The Cerro Gordo mining district. The Pacific Mineralogist 2, no 1:9-11 (1935)

Webb, Robert W.—Cont.

36 Guide to the geology of the route from Los Angeles to Bakersfield, California. The Pacific Mineralogist 3, no 2:6-7 (1936)

36a Kern Canyon fault, southern Sierra Nevada. J G 44:631-638 (1936) (*abst*) G Zentralbl, Abt A, Bd 58:156 (1936) (*abst*) Rv Geol et Sci conn 16:446 (1936)

Webber, Irma E.

33 Woods from the Ricardo Pliocene of Last Chance Gulch, California. Carnegie Inst, Wash, Pub 412:113-134, 5 pls (1933) (*abst*) N Jb 1934, Ref 3:358 (1934) (*abst*) Pale Zentralbl 4:306 (1934) Biol Absts 8:1302, entry 12077 (1934)

Weir, Walter W.

32 Soil erosion in California: its prevention and control. Cal Univ, Coll Agri, B 538:46 pp, 39 figs (1932) (*abst*) Rv Geol et Sci conn 14:293 (1934) (*abst*) G Zentralbl 51:445 (1934)

36 (and **Storie, R. Earl**) A rating of California soils. Agr Exp Station, Berkeley, Cal, B 599:157 pp, tables, maps (1936)

Weirich, T. E.

32 (Discussion to article on petroleum production in Oklahoma during 1931) *In* Petroleum Dev and Tech, Am I M Eng, Tr 1932:170 (1932)

Wells, Arthur E.

31 Sulphur, pyrite and sulphuric acid. Min Ind 39:559-571, California 565 (1931)... 40:509-519, California 515 (1932)... 41:485-496, California 492 (1933)... 42:538-548, California 544 (1934)... 43:538-548, California 538, 544 (1935)

Welsch, O. D.

30 Testing ores for recovery of gold content. M J Ariz 18, no 19:11 (1930)

West, H. E.

28 New attempt to develop Temescal tin deposit in southern California. (*abst*) *In* Ore deposits of the western states. Am I M Eng, Pub 1933:561 (1933)

Westsmith, J. N.

30 Lower horizons in Kettleman Hills proved by general well. Nat Pet News 22, no 26:52-53 (1930)

30a Superior well proves eleven mile axis for Kettleman Hills field. Nat Pet News 22, no 37:43-44 (1930)

30b Shallow sand at Playa del Rey adds to field's prospects. Nat Pet News 22, no 45:43-44 (1930)

30c Temblor formation found in deep test in Lost Hills field. Nat Pet News 22, no 50:44-45 (1930)

Westsmith, J. N.—Cont.

31 Ocean floor surveys show off-shore extension to Rincon field. Inst Pet Tech, J 8:243-244, 2 figs (1931) (*abst*) Eng Index 1931:1036 (1931)

31a New deep sand in Ventura Avenue field discovered at 8800 feet. Nat Pet News 23:44-45 (1931)

Wetmore, Alexander

30 Fossil bird remains from the Temblor formation near Bakersfield, California. Cal Ac Sc, Pr (4) 19:85-93, 7 figs (1930) (*abst*) G Zentralbl 43:282 (1930-1931) (*abst*) N Jb 1933, Ref 3:208 (1933)

31 The fossil birds of North America. Check-list of North American birds. American Ornithologist's Union 1931:401-472 (1931)

33 Development of our knowledge of fossil birds. Fifty Years Progress of American Ornithology 1883-1933:231-239 (1933)

Weymouth, A. Allen

31 (and **Barbat,** W. F.) Stratigraphy of *Borophagus* littoralis locality, California. Cal Univ Dp G, B 23, no 3:25-36, 2 figs, 2 pls (1931) (*abst*) Pale Zentralbl 2:266-267 (1933) (*abst*) G Zentralbl, Abt A 48:253 (1932)

Whaley, W. C.

33 Reconditioning of oil wells in California. Petroleum World (London) 30:320-323 (1933)

Wheeler, Harry Eugene

32 Fusulinids of McCloud and Nosoni formations (Shasta County, California) (*abst*) Pan Am G 58:149 (1932) (*abst*) G Soc Am, B 44:218 (1933) (*abst*) Pale Zentralbl 4:212 (1934)

34 The Carboniferous-Permian dilemma. J G 42:62-70 (1934) (*abst*) G Soc Am, Pr 1933:302 (1933) (*abst*) Pale Zentralbl 5:84 (1934)

35 New trilobite species from the anthracolithic of northern California. San Diego Soc N H, Tr 8:47-58, 1 pl (1935) (*abst*) G Soc Am, Pr 1934:386-387 (1935) (*abst*) Pan Am G 62:71 (1934)

35a The fauna and correlation of the McCloud limestone of northern California. Stanford Univ, Abstracts Dissertations 10:83-85 (1934-1935)

36 Stratigraphy and fauna of the McCloud limestone. (*abst*) G Soc Am, Pr 1935:409 (1936) (*abst*) Pan Am G 63:370 (1935)

Whelden, Roy M.

30 Diatomaceous earth or diatomite. Rocks & Min 5:43 (1930)

White, A. G.

36 (and **Hopkins,** G. R.; and **Breakey,** H. A.) Crude petroleum and petroleum products. U S B M, Min Yb 1936:667-723, California 668, 674, 678, 681, 686, 687, 691, 693, 695, 711, 713 (1936)

Whitfield, James Edward

10 Analyses of borates. U S G S, B 419:300 (1910)

Wicks, Frank R.

31 Crystalline talc. Operations in California of the Pacific Talc Co. Cal Dp Nat Res, Div Mines, Mining in California, St Mineralogist's Rp 27:100-104 (1931) (*abst*) N Jb 1931, Ref 2:552-553 (1931) Eng M World 2:37-39 (1931)

Wickson, Gladys G.

31 (and **Bryan,** K.) The W. Penck method of analysis in southern California. Zs Geomorph 6:287-291 (1931)

33 New Miocene mollusks from California. (*abst*) N Jb 1933, Ref 3:152 (1933) (*abst*) G Zentralbl 43:268-270 (1930-1931) (*abst*) Pale Zentralbl 5:118 (1934) Biol Absts 8:2088-2089, entry 18903 (1934)

34 Some previously unpublished figures of type mollusks from California. (*abst*) N Jb 1934, Ref 3:282-283 (1934)

Wiebe, Walter A., ver

30 Oil fields of the United States. 629 pp, 230 illus (1930) (*abst*) G Zentralbl 43:168 (1930-1931) (*Review*) M Mag 43:90-91 (1930) (*ann*) An Bib Ec G 3:143-144 (1931)

32 Present distribution and thickness of Paleozoic systems. G Soc Am, B 43:495-540, California 500, 506, 510, 512, 513, 514, 517, 518, 519, 525-526, 527, 531, 532, 533, 538, maps (1932)

33 Present distribution and thickness of Mesozoic systems. G Soc Am, B 44:827-864, California 830, 831, 832, 838, 839, 840-843, 846, 847, 852, 853, 854, 863 (1933)

Wiel, S. C.

05 Ancient channels at Gibsonville, California. M Sc Press 91:73 (1905)

Wilhelm, C. J.

35 (**Thorne,** H. M.; and **Schmidt,** L.) Disposing of oil field brines. Petroleum World (London) 32:103-106, California 105 (1935)

Wilhelm, Salomon C.

32 Epeirophorese. III. Teil. Die vordiluvialen Eiszeiten. A. Die Eiszeiten des Tertiars und Mesozoicums. (*abst*) N Jb 1932, Ref 2:169-170 (1932)

Wilhelm, V. H.

32 Developments in the California petroleum industry during 1931. *In* Petroleum Dev and Tech, Am I M Eng, Tr 1932:182-195 (1932)

33 (**Davis**, E. L.; and **Clark**, W. A.) Characteristics of edge-water encroachments in California fields. Oil Weekly 71, no 4:13-16 (1933) (*ann*) An Bib Ec G 6:294 (1934) M Metal 14:423-425 (1933) (*abst*) Rv Geol et Sci conn 15:354 (1935)

33a (and **Miller**, H. W.) Developments in the California petroleum industry during 1932. *In* Petroleum Dev and Tech, Am I M Eng, Tr 1933:345-351 (1933)... 1933, Am I M Eng, Tr 1934:182-197 (1934)

34 (and **Miller**, H. W.) Developments in the California oil industry during the year 1933. Am I M Eng, Tr 107:182-197 (1934)

35 Development in the California oil industry during the year 1934. Am I M Eng, Tr 114:257-269 (1935)

36a Petroleum development and production in the future. M Metal 17:343-346 (1936)

Williams, Howell

30 The volcanic domes of Lassen Peak and vicinity, California. (*abst*) Zs Vulkan 13:223 (1930)

30a Geology of the Marysville Buttes, California. (*abst*) Zs Vulkan 13:224 (1930) (*Review*) Sc Progress 24:581 (1929-1930)

30b A recent volcanic eruption near Lassen Peak, California. (*abst*) Zs Vulkan 12:268 (1929-1930) (*abst*) Peterman's Mitt 76:275-276 (1930)

31 The dacites of Lassen Peak and vicinity, California, and their basic inclusions. Am J Sc (5) 22:385-403, 7 figs (1931) (*abst*) G Zentralbl 46:406 (1932) Zs Vulkan 15:224 (1933) (*Review*) Sc Progress 27:439 (1933) Chem Absts 26: 943 (1932)

31a The history and character of volcanic domes. Cal Univ, Dp G, B 21:51-146, 37 figs (1932) (*abst*) Zs Vulkan 15: 224 (1933) (*abst*) Eng Index 1932:616 (1932)

32 Geology of the Lassen Volcanic National Park, California. Cal Univ, Dp G, B 21:195-385, 2 maps, 64 figs (1932) (*Review*) Sc Progress 29:120 (1934) Zs Vulkan 16:69-73 (1934) (*abst*) N Jb 1934, Ref 2:563 (1934)

32a Mount Shasta, a Cascade volcano. J G 40:417-430 (1932) (*abst*) Rv Geol et Sci conn 13:20 (1932) N Jb 1934, Ref 2:63-64 (1934) (*abst*) G Zentralbl, Abt A 48: 282 (1932) (*abst*) Zs Vulkan 15:224 (1933) (*Review*) Sc Progress 29:121 (1934)

33 Mount Thielsen, a dissected Cascade volcano. Cal Univ, Dp G Sc, B 23:195-213, 13 figs (incl geol map) (1933)

Williams, Howell—Cont.

34 Mount Shasta, California. Zs Vulkan 15:225-253 pls 15-21, 1 fig (1934)

35 (and **Evans**, R. D.) The radium content of lavas from Lassen Volcanic National Park, California. Am J Sc (5) 29:441-452 (1935)

Williams, I. A.

31 (with **Berkey**, C. P.; **Louderback**, G. D.; **Hinderlider**, M. C.; and **Savage**, J. L.) Report of consulting board on safety of the proposed Pine Canyon dam, Los Angeles County. Cal Dp Pub Works, Div Water Resources May:22 pp. 8 pls, 1 map (1931)

Willis, Bailey

25 (and **Dewell**, H. D.) Earthquake damage to buildings. Seism Soc Am, B 15:282-301, 15 pls (1925)

26 Seismicity of Chile and California. Third Pan-Pacific Sc Cong Tokyo, Pr 1: 389-394 (1926)

Wilmarth, Grace

31 Names and definitions of the geologic units of California. U S G S, B 826:97 pp (1931) (*abst*) Rv Geol et Sci conn 13:562-563 (1933) (*abst*) G Zentralbl 45:386 (1931)

Wilson, C. H.

32 (and **Jakosky**, J. J.) Use of geophysics in placer mining. M J Ariz 16, no 14:3-4, 29 (1932) (*abst*) Rv Geol et Sci conn 13:416 (1933) (*abst*) Eng Index 1932:617 (1932) (*abst*) G Zentralbl 51:485-486 (1934)

34 (and **Jakosky**, J. J.) Geophysical studies in placer and water supply problems. Am I M Eng, Tech Pub 515:18 pp, pls (1934) Eng M J 1935:71-74 (1934) (*abst*) Eng Index 1934:519 (1934)

36 (and **Jakosky**, J. J.) Electrical mapping of oil structures. M Metal 17:231-237 (1936)

Wilson, James T.

35 (and **Byerly**, P.) The Richmond quarry blast of August 16, 1934. Seism Soc Am, B 25:259-269 (1935)

35a (and **Byerly**, P.) The central California earthquakes of May 16, 1933, and June 7, 1934. Seism Soc Am, B 25:223-247 (1935)

35b (and **Byerly**, P.) Earthquakes in northern California and the registration of earthquakes at Berkeley, Mount Hamilton, Palo Alto, and San Francisco from April 1, 1933 to September 30, 1933. Cal Univ Seism Sta, B 4:1-73 (1935)

35c (and **Byerly**, P.) Northern California earthquakes, April 1, 1933 to March

Wilson, James T.—Cont.

31, 1934. Seism Soc Am, B 25:269-273 (1935)

35d (and **Byerly, P.**) Earthquakes in northern California and the registration of earthquakes at Berkeley, Mount Hamilton, Palo Alto, San Francisco, Ferndale from October 1, 1933 to March 31, 1934. Cal Univ Sta, B 4:75-165 (1935)... April 1, 1934 to September 30, B 4:167-243 (1936)... October 1, 1934 to March 31, 1935, B 4:245-338 (1936)

36 (and **Annis, W.**) Earthquakes in northern California and the registration of earthquakes at Berkeley, Mount Hamilton, Palo Alto, San Francisco, Ferndale from April 1, 1935 to June 30, 1935, Cal Univ Seism Sta, B 5:1-38 (1936)

Wilson, Leslie E.

35 Miocene marine mammals from the Bakersfield region, California. Peabody Mus N H, B 4:140 pp, figs (1935)

Wilson, Robert W.

32 Pleistocene rodent fauna from Carpinteria asphalt deposits. (*abst*) Pan Am G 58:150 (1932) G Soc Am, B 44:219-220 (1933)

32a *Cosomys*, a new genus of vole from the Pliocene of California. J Mammalogy 13:150-154, 1 pl (1932)

33 Pleistocene mammalian fauna from the Carpinteria asphalt. Carnegie Inst, Wash, Pub Paleontology 440:59-76 (1934)

34 New fauna from the Sespe of Las Posas Hills. (*abst*) Pan Am G 62:69 (1934) (*abst*) G Soc Am, Pr 1934:384 (1935)

35 Cricetine-like rodents from the Sespe Eocene of California. Nat Ac Sc, Pr 21: 26-32 (1935) (*abst*) Pale Zentralbl 4:235 (1934)

35a Two rodents and a lagomorph from the Sespe of the Las Posas Hills, California. Carnegie Inst, Wash, Pub 453: 11-17, 1 pl, 1 fig (1935)

35b *Simimys*, a new name to replace *Eumysops Wilson*, Pre-occupied—a correction. Nat Ac Sc, Pr 21:179-180 (1935)

Wilson, W. B.

34 Proposed classification of oil and gas reservoirs. *In* Problems of petroleum geology. Am As Petroleum G, Sidney Powers Memorial Volume:433-445 (1934)

Wisker, A. L.

36 The gold-bearing veins of Meadow Lake District, Nevada County. Cal Dp Nat Res, Div Mines, M G, J, St Mineralogist's Rp 32:189-204 (1936)

Woelflin, William

35 Centrifuge tests on cut oil. Petroleum World (London) 32:263-264 (1935)

Wolff, John Eliot

30 Dumortierite from Imperial County, California. (*abst*) Am Mineralogist 15: 119 (1930) Min Absts 5:43-44 (1932)

31 Reconnaissance of the southern Panamints and of Ashford Canyon in the Black Mountains. (*abst*) G Soc Am, B 43:225 (1932) (*abst*) Pan Am G 55:360 (1931)

Wood, A. E.

36 Cuyama Tertiary fauna of California. (*abst*) G Soc Am, Pr 1935:395 (1936)

36a Fossil heteromyid rodents in the collections of the University of California. Am J Sc 32 (5):112-119 (1936) (*abst*) G Soc Am, Pr 1935:401 (1936)

Wood, Harry Oscar

31 (and **Buwalda, J. P.**) Horizontal displacement along the San Andreas fault in the Carrizo Plain, California. (*abst*) G Soc Am, B 42:298-299 (1931) (*abst*) Pan Am G 54:75 (1930)

31a (with **Richter, C. F.**) Recent earthquakes near Whittier, California. Seism Soc Am, B 21:183-203 (1931) (*ann*) An Bib Ec G 5:191 (1932) (*abst*) G Zentralbl, Abt A 48:496 (1933)

31b (and **Richter, C. F.**) A study of blasting recorded in southern California. Seism Soc Am, B 21:28-46 (1931) (*abst*) N Jb 1931, Ref 2:661 (1931) (*abst*) G Zentralbl, Abt A 48:477 (1933)

32 (**Buwalda, J. P.; and Gutenberg, B.**) Experiments testing seismographic methods for determining crustal structure. (*abst*) Pan Am G 58:65-66 (1932) Seism Soc Am, B 22:185-242 (1932)

32a (**Gutenberg, B.; and Richter, D. F.**) The earthquake in Santa Monica Bay, California, on August 30, 1930. Seism Soc Am, B 22:138-154, 1 fig, 2 pls (1932) (*abst*) N Jb 1933, Ref 2:573-574 (1933)

33 The Long Beach earthquake. Science n s 78:147-148 (1933)

33a Note on the Long Beach earthquake. Science n s 78:281-282 (1933)

33b "Apparent" intensity and surface geology. Nat Res Council, B 90:67-82 (1933)

33c Earthquake investigation in the field. Nat Res Council, B 90:41-66 (1933)

33d Preliminary report on the Long Beach earthquake of March 10, 1933. Seism Soc Am, B 23:43-56 (1933) (*abst*) Rv Geol et Sci conn 13:554-555 (1933) (*abst*) N Jb 1934, Ref 2:550 (1934) (*abst*) Nature 133:108 (1934)

33e (and **Richter, C. F.**) A second study of blasting recorded in southern California. Seism Soc Am, B 23:95-110, 2 pls (1933) (*abst*) Rv Geol et Sci conn 13:620

Wood, Harry Oscar—Cont.

(1933) (*abst*) N Jb 1934, Ref 2:550 (1934) (*abst*) G Zentralbl, Abt A 47:177 (1932)

34 Earthquakes in California. Sc Mo 39:323-344 (1934)

35 (**Allen,** M. W.; and **Heck,** N. H.) Destructive and near-destructive earthquakes in California and western Nevada, 1769-1933. U S G S, Sp Pub 191:24 pp (1934)

Woodford, A. O.

26 The Catalina metamorphic facies of Franciscan series. (*abst*) N Jb 1926, Ref 2, Abt B:199 (1926)

26a The San Onofre breccia: its nature and origin. (*abst*) N Jb 1926, Ref 2, Abt B:198 (1926)

32 (and **Laudermilk,** J. D.) Rilled limestone. (*abst*) G Soc Am, B 43:227 (1932)

32a (and **Laudermilk,** J. D.) Concerning Rillensteine. Am J Sc (5) 23:135-154, figs (1932)

32b (and **Laudermilk,** J. D.) Soda-rich anthophyllite asbestos from Trinity County, California. Min Absts 5:45 (1932)

33 Clay minerals of California soils. (*abst*) G Soc Am, Pr 1933 1:313-314 (1934)

33a (**Kelley,** W. P.; and **Brown,** S. M.) Clay minerals of California soils. (*abst*) Pan Am G 59:315-316 (1933)

33b (and **Laudermilk,** J. D.) California occurrence of montmorillonite after feldspar. (*abst*) Pan Am G 59:315 (1933) (*abst*) G Soc Am, Pr 1933 1:313 (1934) (*abst*) Rv Geol et Sci conn 14:375 (1933-1934)

33c (and **Taylor,** E.) Longitudinal profiles of streams. (*abst*) G Soc Am, Pr 1933 1:307 (1934)

34 (and **Laudermilk,** J. D.) Secondary montmorillonite in a California pegmatite. Am Miner J 19:260-267 (1934) (*abst*) N Jb 1934, Ref 1:464 (1934) (*ann*) An Bib Ec G 7:218 (1934)

35 (and **Laudermilk,** J. D.) Black iron sulphide in California crystalline limestone. (*abst*) Pan Am G 63:320 (1935) (*abst*) G Soc Am 1935:342 (1936)

35a Rhomboid ripple mark. Am J Sc (5) 29:518-525, figs (1935)

36 (and **Foshag,** W. F.) Bentonitic magnesian clay-mineral from California. Am Mineralogist 21:238-244 (1936) (*abst*) Rv Geol et Sci conn 16:435-436 (1936)

Woodhouse, C. D.

31 (with **Jeffery,** J. A.) Note on a deposit of andalusite in Mono County, California: its occurrence and technical importance. Cal Dp Nat Res, Div Mines, Mining in California, St Mineralogist's Rp 27:459-464 (1931) (*ann*) An Bib Ec G 4:271 (1931)

Woodhouse, C. D.—Cont.

32 (and **Jeffery,** J. A.) Mining andalusite in Mono County, California. M J Ariz 15, no 16:5-6, 43-44 (1932)

34 A new occurrence of montroydite in California. Am Mineralogist 19:603-604 (1934) (*ann*) An Bib Ec G 7:218 (1934)

36 Change them every 10,000 miles. The Mineralogist 4, no 3:3-4, 37-38 (1936)

Woodring, Wendell Phillips

30 Tertiary deposits bordering the Simi Valley, California. (*abst*) G Zentralbl 43:267 (1930-1931) (*abst*) G Soc Am, B 42:299 (1931)

31 A Miocene Haliotis from southern California. J Paleontology 5:34-39, 1 pl (1931) (*abst*) Pale Zentralbl 1:370 (1932)

31a Age of the orbitoid-bearing Eocene limestone and *Turritella variata* zone of the western Santa Ynez Range, California. San Diego Soc N H, Tr 6:371-388 (1931) (*abst*) G Zentralbl, Abt A 48:70 (1932) (*abst*) Pale Zentralbl 2:292 (1933) Biol Absts 8:526, entry 4671 (1934)

31b Upper Eocene orbitoidal foraminifera from the Santa Ynez Range, California. (*abst*) G Soc Am, B 42:370 (1931) (*abst*) Pale Zentralbl 2:292 (1933)

32 Miocene mollusk of the genus *Haliotis* from the Temblor Range, California. U S Nat Mus, Pr 81, art 15, 4 pp, 1 pl (1932) (*abst*) Rv Geol et Sci conn 13:443 (1933) (*abst*) Pale Zentralbl 4:276 (1934)

32a (**Roundy,** P. V.; and **Farnsworth,** H. R.) Geology and oil resources of the Elk Hills, California (including Naval Petroleum Reserve No. 1). U S G S, B 835:82 pp, 8 figs, 22 pls (1932) (*abst*) Rv Geol et Sci conn 13:367, 513 (1933) (*abst*) Eng Index 1933:818 (1933)

32b Distribution and age of the marine Tertiary deposits of the Colorado Desert. Carnegie Inst, Wash, Contr to Paleontology, Pub 418:1-425, 1 fig (1931) (*abst*) Rv Geol et Sci conn 13:339-340 (1933)

32c (and **Kew,** W. S. W.) Tertiary and Pleistocene deposits of the San Pedro Hills, California. (*abst*) Wash Ac Sc, J 22, no 2:39-40 (1932)

32d San Pedro hills. XVI. Int G Cong, Guide Book 15:34-40, figs (1932)

33 Pliocene deposits north of Simi Valley, California. (*abst*) N Jb 1933, Ref 3:560-561 (1933) (*abst*) Pale Zentralbl 3: 318 (1933) Biol Absts 7:454, entry 4333 (1933)

35 (**Bramlette,** M. N.; and **Kleinpell,** R. M.) Miocene stratigraphy and paleontology of the Palos Verdes Hills. Am As Petroleum G, B 19:1842 (1935)

35a Fossils from marine Pleistocene terraces of the San Pedro Hills, Califor-

Woodring, Wendell Phillips—Cont.

nia. Am J Sc 29:292-305, fig (1935) (abst) Eng Index 1935:501 (1935)

35b Pliocene viviparoid calcareous opercula from Kettleman Hills. (abst) Pan Am G 63:375 (1935)

36 New Miocene fauna from the California Coast Ranges. (abst) G Soc Am, Pr 1935:366 (1936)

36a (Kleinpell, R. M.; and Bramlette, M. N.) Miocene stratigraphy and paleontology of Palos Verdes Hills, California. Am As Petroleum G, B 20:125-149, figs (1936) (abst) G Zentralbl, Abt A, Bd 57: 51-52 (1936) (abst) Rv Geol et Sci conn 16:378 (1936)

Woodworth, S. E.

31 Milling methods and costs at the Argonaut Mill, Jackson, California. U S B M, Inf Circ 6476:12 pp, 3 figs (1931) (abst) U S B M, List Pub 1910-1932:149 (1933) (abst) Eng Index 1931:676 (1931)

Woolf, J. A.

31 (and Leaver, E. S.) Re-treatment of Mother Lode (California) carbonaceous slime tailing. U S B M, Tech P 481: 20 pp (1931)

Wrather, W. E. (Ed.)

34 (and Lahee, F. H. (Ed.)) Problems of petroleum geology: a symposium. Am As Petroleum G, Sidney Powers Memorial Volume (1934) (Review) Ec G B 30:194-196 (1935) Inst Pet Tech, J 20:1113-1115 (1934)

Wright, C. W.

32 (and Meyer, H. M.) Lead. U S B M, Min Res, Min Yb 1932-1933:53-66, California 58 (1933)

Wright, W. Quinby

34 (and Jenkins, Olaf P.) California's gold-bearing Tertiary channels. Eng M J 135:497-502 (1934) (abst) G Zentralbl, Abt A, Bd 55:457 (1935)

Wu, C. C.

31 (and Trask, P. D.) Analyses of oil and gas from distillation of recent sediments. (abst) G Zentralbl 43:98 (1930) (ann) An Bib Ec G 3:135 (1931) (abst) N Jb 1931, Ref 2:513 (1931)

Yaeckel, M. P.

34 Benitoite, California's exclusive gem. Oregon Mineralogist 2:26-27 (1934)

Young, A. A.

30 (with Blaney, H. F.; and Taylor, C. A.) Rainfall penetration and consumptive use of water in Santa Ana River valley and coastal plain, California. Cal Dp Nat Res, B 33:158 pp (1930) (ann) An Bib Ec G 4:333 (1931)

Young, G. J.

31 Gold-ore mining and milling. Eng M J 132:195-199, 4 figs (1931) (abst) Eng Index 1931:673 (1931)

34 (and Romanowitz, C. M.) Gold-dredging. Eng M J 135:486-490 (1934)

Young, W. H.

31 (with Tryon, F. G.; and Corse, J. M.) Fuel briquets. U S B M, Min Res, Calendar Year 1931:61-71, California 65, 67 (1933)

31a (Mann, L.; Tryon, F. G.; Berquist, F. E.; and Bennit, H. L.) Coal. U S B M, Min Res, Calendar Year 1931:415-510, California 426, 472 (1933)

32 (and Corse, J. M.) Fuel briquets. U S B M, Min Yb 1932-1933, Calendar Year 1932:451-458, California 456 (1933)

32a (Mann, L.; Bennit, H. L.; and Tryon, F. G.) Coal. U S B M, Min Yb, Stat App, Calendar Year 1932:373-454, California 376, 423 (1934)... 1933:281-360, California 286, 323 (1935)

33 (and Clark, J. B.) Fuel briquets. U S B M, Min Yb 1934, Calendar Year 1933:645-652, California 648, 651 (1934)... 1935:Calendar Year 1934:711-718, California 715, 717 (1935)

Youngman, E. P.

31 Zircon. U S B M, Inf Circ 6465:20 pp, 1 fig (1931) (abst) Eng Index 1931: 1268 (1931)

31a Zirconium. U S B M, Inf Circ 6456:1-31, California 5-6 (1931)

Zadach, S.

33 Placer mining in San Gabriel Canyon. M J Ariz 17, no 3:3 (1933)

Zepharovich, V. von

85 Kallait pseudomorph nach Apatit aus Californien. Zs Kryst 10:240-251 (1885)

PART II

INDEX FOR BIBLIOGRAPHY OF CALIFORNIA

.The numbers in bold-faced type, set in parentheses, refer to pages in Part I.)

Abrasive: (22) Eardley-Wilmot, V. L. 31
  materials: (8) Bowles, O. 32, 34a; (19) Davis, A. E. 30, 34, 35, 36; (32) Hatmaker, P. 30; (41) Johnson, B. L. 36a
Acila, nuculid pelecypod: (67) Schenck, H. G. 35c, 36c
Adamite, Chloride Cliff: (55) Murdoch, J. 36b
Adelaida quadrangle: (76) Taliaferro, N. L. 35
  geology: (72) Stanton, W. L. 31
Afton, Basin, Mojave desert, physiographic history: (23) Ellsworth, E. W. 33
Agate beds, Mint Canyon: (58) Patton, J. W. 36
Alamitos Heights oil field: (34) Hight, W. 33
Albitization, Inyo Range: (3) Anderson, G. H. 35
Algae, Eocene, Santa Barbara County: (38) Howe, M. A. 33; (56) Nelson, R. N. 31; (67) Schenck, H. G. 31a
Alkaline lakes supply soda: (35) Hirschkind, W. 31
Alleghany district, gold quartz veins: (25) Ferguson, H. G. 30, 32; (28) Gannett, R. W. 32.
Alluvial fans, Cucamonga district: (23) Eckis, R. 31
Alpine County: (47) Logan, C. A. 30
  land forms: (65) Russell, R. J. 33
Amador County, auriferous gravels, dredging: (58) Patmon, C. G. 32
  Central-Eureka mine: (71) Spiers, J. 31; (78) U. S. Bureau of Mines 31
Amazon stone: (1) Aitkens, I. 31
Amber: (55) Murdoch, J. 34
American River, geology of damsites: (25) Forbes, H. 29
Ammonite, Barroisiceras, Cretaceous: (62) Reeside, J. B. 32
  Jurassic, Coast Ranges: (14) Crickmay, C. H. 32
Ammonoids, Triassic: (70) Smith, J. P. 32
Amphistegina californica: (67) Schenck, H. G. 30
Amynodont skull, Sespe: (73) Stock, C. 33c
Analcite diabase: (76) Taliaferro, N. L. 31
Analysis, W. Penck method in southern California: (10) Bryon, K. 31b; (82) Wickson, G. G. 31
Anauxite, volcanic rocks: (64) Rogers, A. F. 32
Anchitherine horses, Merychippus zone, Coalinga: (7) Bode, F. D. 34a
Ancient channels at Gibsonville: (82) Wiel, S. C. 05
  formation, Klamath Mountains: (35) Hinds, N. E. A. 31

Ancient channels—Cont.
  rivers, gravel channels: (7) Skidmore, W. A. 85
Andalusite: (22) Dunn, J. A. 33; (46) Lilley, E. R. 36; (55) Murdock, J. 36a; (65) Sampson, R. J. 31; (77) Tucker, W. B. 31a
  Mono County: (39) Jeffery, J. A. 31, 32; (85) Woodhouse, C. D. 31, 32
  White Mountain: (43) Kerr, P. F. 32
Anorthite: (53) Miller, F. S. 35a
Anorthosite: (53) Miller, W. J. 31a, 32b
Antelope Sphenophalos, Pliocene, distribution: (26) Furlong, E. L. 31
Anthophyllite asbestos, Trinity County: (45) Laudermilk, J. D. 30; (85) Woodford, A. O. 30, 36
Antimony deposits: (68) Schrader, F. C. 33
Antioch quadrangle, Domengine and Markley formations: (81) Weaver, C. E. 31
Arcidae, classification: (63) Reinhart, P. W. 35
  Tertiary: (63) Reinhart, P. W. 33
Arenaceous Temblor foraminifera: (5) Barbat, W. F. 32b; (19) Cushman, J. A. 32
Argonaut, milling: (42) Jullum, H. 32
  mine: (41) Josephson, W. G. 32; (79) Vandenburg, W. O. 31a; (86) Woodworth, S. E. 31
Arid regions, rock floors: (20) Davis, W. M. 32a
Arkose, diatomaceous shales interbedded: (67) Schenck, H. G. 31d
Arroyas: (29) Grant, U. S. 35a; (61) Putman, W. C. 35a
Arroyo Grande oil field: (23) Emmons, W. H. 31
Arsenic: (59) Petar, A. V. 34a; (78) Tyler, P. M. 34
Artiodactyl, Sespe Eocene: (74) Stock, C. 36a
Artiodactyla, Sespe, Las Posas Hills: (74) Stock, C. 35g
Asbestos: (8) Bowles, O. 32c, 32e, 34; (45) Laudermilk, J. D. 30; (74) Stoddard, B. H. 32a; (85) Woodford, A. O. 32b
Asphalt: (38) Hubbard, P. 31; (38) Huttle, J. B. 34a; (79) Vander Hoof, V. L. 34, 35
  production, 1933: (59) Petroleum World 34f
  Rancho La Brea: (73) Stock, C. 32h
  related bitumes: (62) Redfield, A. H. 32
Astrodapsis, antiselli zone: (15) Clark, B. L. 31d
  revision of species: (63) Richards, G. L. 35
Aturia, cephalopods of North America: (67) Schenck, H. G. 31

Augelite, Mono County: (46) Lemmon, D. W. 35

Auk, Lucas: (53) Miller, L. 33

Auluroid, Ordovician: (60) Phleger, F. B. 36

Auriferous gravels, age: (14) Chaney, R. W. 31a, 33

Amador County: (58) Patmon, C. G. 32
 Channels: (54) Mining and Scientific Press 80
 dredging: (58) Patmon, C. G. 32
 floras: (48) MacGinitie, H. 33
 fossils, near Columbia: (34) Louderback, G. D. 34
 Sierra Nevada, fossil plants: (14) Chaney, R. W. 32
 Tertiary mammals: (51) Merriam, J. C. 33; (73, 74) Stock, C. 33h

Avifaunas, Pleistocene, (53) Miller, L. H. 32, 35

Bacteria, petroleum: (77) Thorpe, W. H. 32

Bakersfield, oil fields: (16) Clute, W. S. 36
 petroleum: (16) Clements, T. 36

Barber's Hill deep sand development: (41) Judson, S. A. 30; (56) Murphy P. C. 30

Barite: (9) Bradley, W. W. 30a; (18) Cornthwaite, M. A. 36; (25) Fitch A. A. 31; (41) Johnson, B. L. 36; (46) Lilley, E. R. 36; (66) Santmyers, R. M. 32a; (74) Stoddard, B. H. 32b
 microcrystals from Barstow: (37) Howard, A. D. 32

Barium: (32) Hardy, C. 31
 products: (18) Cornthwaite, M. A. 36; (41) Johnson, B. L. 36; (66) Santmyers, R. M. 32a; (74) Stoddard, B. H. 32b

Barrancos: (29) Grant, U. S. 35a; (61) Putman, W. C. 35

Barrelian series, eastern California: (43) Keyes, C. R. 31

Barroisiceras, Cretaceous ammonite: (62) Reeside, J. B. 32

Barstow, beds, lagomorphs and rodents: (31) Hall, E. R. 34
 Miocene, Soricidae: (73) Stirton, R. A. 32

Basalt: (8) Bowles, O. 34b

Basin, Range, Pleistocene glaciation: (7) Blackwelder, E. 31e, 34
 lakes: (6) Blackwelder, E. 31c
 structures: (28) Gilbert, G. K. 31; (65) Russell, R. J. 30
 earthquake of December 20, 1932: (13) Callaghan, E. 34; (28) Gianella, V. P. 34
 talus slopes: (7) Blackwelder, E. 34a
 south coastal investigation: (12, 13) California Department of Public Works 32, 33a, 34f, 35b

Bassarisk, Tertiary: (31) Hall, E. R. 30

Batholith, Grass Valley, structure: (41) Johnston W. D. 32b
 Sierra Nevada: (16) Cloos, E. 32, 33, 33a, 33b

Bavenite, beryllium mineral: (24) Fairchild, J. G. 32; (67) Schaller, W. T. 32

Beach deposits, stratification: (76) Thompson, W. O. 33

Beach deposits—Cont.
 placers, Oregon coast: (58) Pardee, J. T. 34
 structure: (76) Thompson, W. O. 35

Beavers, Tertiary: (73) Stirton, R. A. 35

Bedding, Monterey Rocks: (9) Bramlette, M. N. 33

Bed-rock complex: (76) Taliaferro, N. L. 32a, 32b

Belridge oil field: (23) Emmons, W. H. 31

Ben Lomond Mountain, geology: (25) Fitch, A. A. 31a

Benitoite: (47) Louderback, G. D. 09; (86) Yaeckel, M. P. 34

Bentonite: (35) Hill, H. R. 34; (43) Kerr, P. F. 31; (46) Lilley, E. R. 36; (79) Van Amringe, E. V. 34

Bentonitic magnesian clay-mineral: (26) Foshag, W. F. 36; (85) Woodford, A. O. 36

Berkeley, earthquake registration: (11, 12) Byerly, P. 27, 27a, 28, 28a, 30, 30a, 31c, 33d, 33e, 35, 35a, 35d, 36, 36a; (3) Annis, W. 36, 36a; (22) Dyk, R. 28, 30, 30a; (37) Hoskins, E. E. 36; (41) Jones, A. E. 27; (71) Sparks, N. R. 32, 33a, 33b; (83, 84) Wilson, J. T. 35b, 35d, 36
 Hills: (15) Clark, B. L. 33
 Pliocene sequence: (15) Clark, B. L. 32a, 33a

Beryl: (59) Petar, A. V. 31

Beryllium: (20, 21) Déribéré M. 34; (59) Petar, A. V. 31
 bavenite: (24) Fairchild, J. G. 32; (67) Schaller, W. T. 32

Bibliography, foraminifera: (76) Thalman, H. E. 35
 geology, mineral resources of California: (69) Shedd, S. 31
 North American: (57) Nickles, J. M. 31; (76) Thom, E. M. 35
 placer mining: (26) Franke, H. A. 32
 stones, ventifacts: (10) Bryan, K. 31a

Biostratigraphic classification, Miocene: (44) Kleinpell, R. M. 34

Birds, fossil: (82) Wetmore, A. 31, 33
 horizons: (53) Miller, L. 35a
 Pleistocene of Carpinteria: (53) Miller, L. H. 31a
 remains, Pliocene: (37) Howard, H. 31; (53) Miller, L. H. 31
 Pleistocene: (53) Miller, L. H. 32c

Black Mountains: (84) Wolff, J. E. 31
 sand deposits, northern California: (37) Horner, R. R. 18
 flour gold: (42) Kellogg, A. E. 31

Blasting study, southern California: (63) Richter, C. F. 31b, 33; (84, 85) Wood, H. O. 31b, 33e

Bolivinas, Tertiary: (1) Adams, B. C. 35; (19) Cushman, J. A. 35b

Booming: (45) Laizure, C. M. 33

Borates: (1) Anonymous 33; (80) Vonson, M. 29
 analyses: (82) Whitfield, J. E. 10
 discovery: (80) Vonson, M. 35
 minerals, Kramer district: (67) Schaller, W. T. 30

Borax: (4) Ayers, W. O. 14; (48) Manning, P. D. V. 30; (13) Calvert, E. L. 34

Geology, economic—Cont.

El Segundo oil field: (61) Powell, E. B. 36

Elizabeth Lake quadrangle: (69) Simpson, E. C. 34

Elk Hills, oil resources: (85) Woodring, W. P. 32a; (65) Roundy, P. V. 32; (24) Farnsworth, H. R. 32

Elsinore quadrangle: (23) Engel, R. 31

geothermics applications: (80) Van Orstrand, C. E. 34a

glauconite: (27) Galliher, E. W. 35b

gold: (40) Jenkins, O. P. 33

historical, difficulty of using cartographic terminology: (44) Kleinpell, R. M. 34c

Inyo Range: (44) Knopf, A. 27

Julian region: (21) Donnelly, M. 34, 34a

Lassen Volcanic National Park: (83) Williams, H. 32

Laurel and Convict Basins, Mono County: (50) Mayo, E. B. 34

Los Angeles: (70) Soper, E. K. 32a; (29) Grant, U. S. 32

to Bakersfield: (81) Webb, R. W. 36

Basin: (36) Hoots, H. W. 32b

late Quaternic: (23) Eckis, R. 35

Lucia quadrangle: (63) Reiche, P. 34

Madera County: (24) Erwin, H. D. 34

Marysville Buttes: (83) Williams, H. 30a

McKittrick oil field: (75) Taff, J. A. 33

middle California: (46) Lawson, A. C. 33

Mokelumne area: (76) Taylor, G. H 31; (64) Robinson, T. W. 31; (72) Stearns, H. T. 31; (76) Thomas, H. E. 32; (27) Gale, H. S. 32a; (60) Piper, A. M. 32a

Mono County: (50) Mayo, E. B. 30

Mount Diablo: (75) Taff, J. A. 35

Mount Jura: (18) Crickmay, C. H. 31

Pinos quadrangle: (22) Dreyer, F. E. 35; (28) Gazin, C. L. 31

Shasta quadrangle: (40) Jenkins, O. P. 32

ore deposits: (10) Butler, B. S. 33, 33a

Panamint Range: (56) Murphy, F. M. 31, 32

silver district: (56) Murphy, F. M. 30

Paso Robles quadrangle: (76) Taliaferro, N. L. 35

Perris block, southern California: (22) Dudley, P. H. 31

petroleum: (23) Emmons, W. H. 31; (45) Lahee, F. H. (Ed.) 34b; (78) Union Oil Company Bulletin 36; (86) Wrather, W. E. (Ed.) 34; (79) U. S. Department of the Interior, Geological Survey 34a

progress: (78) Union Oil Company Bulletin 35

Pit River damsites: (49) Marliave, C. 33

placers, Mojave desert: (38) Hulin, C. D. 35

Plumas County: (2) Anderson, C. A. 31

Potter Creek Cave: (70) Sinclair, W. J. 06

quicksilver deposits: (6) Becker, G. F. 90

Ravenna quadrangle: (68) Sharp, R. P. 35

Geology—Cont.

regional and vertical intensity: (70) Somers, G. B. 30

Riverside area: (4) Bacon, C. S. 33

Sacramento, River, Basin damsites: (26) Forbes, H. 31a

canyon between Cottonwood Creek and Iron Canyon: (49) Marliave, C. 31; (47) Louderback, G. D. 31

Valley: (26) Forbes, H. 31

Salinas quadrangle: (33) Herold, C. L. 35

San Francisco Bay region: (77) Tolman, C. F. 31

San Gabriel Mountains, western: (54) Miller, W. J. 35

San Jacinto quadrangle: (26) Fraser, D. M. 31

San Joaquin River basin damsites: (26) Forbes, H. 31b

Valley: (26) Forbes, H. 31c (36) Hoots, H. W. 31

San Miguel Island:(9) Bremner, C. S. 33

San Simeon quadrangle: (76) Taliaferro, N. L. 35

Santa Ana Mountains: (55) Moore, B. M. 30

Santa Barbara Mesa field: (61) Powell, E. B. 34

Santa Cruz Island: (9) Bremner, C. S. 32; (66) Rand, W. W. 31

Santa Monica Mountains: (36) Hoots, H. W. 31a, 32a

Santa Rosa Island: (55) Moody, G. B. 35a

Sharktooth Hill: (31) Hanna, G. D. 31b

Shasta quadrangle: (3) Averill, C. V. 31

Sierra Nevada: (44) Knopf, A. 27; (49) Matthes, F. E. 33; (53) Miller W. J. 31b

Siskiyou County: (50) Maxson, J. H. 33a

Solano County: (4) Bailey, T. L. 31a

Soledad quadrangle: (57) Nickell, F. A. 30

Southern California: (27) Gale, H. S. 32

earthquakes: (23) Eaton, J. E. 33c

structural geology: (43) King, P. B. 32

submarine: (73) Stockman, L. P. 30

from air: (6) Blackman, E. O. 35

subsurface, Venice: (80) Warner, T. 30

surface (earthquakes): (84) Wood, H. O. 33b

text book: (19) Davies, A. M. 35a

violated: (13) California Oil World 33

Tertiary faunas: (19) Davies, A. M. 35

Transverse Ranges: (36) Hollister, J. S. 35a; (62) Reed, R. D. 35b

Trinity River; Fairview damsite: (47) Louderback, G. D. 33

Vacaville-Rumsey Hills area: (43) Kirby, J. M. 35

valley fill: (23) Eckis, R. 34

Ventura quadrangle: (43) Kerr, P. F. 31; (55) Muller, S. 31; (67) Schenck, H. G. 31b

Weaverville district: (35) Hinds, N. E. A. 34a

White Mountain quadrangle: (3) Anderson, G. H. 31

Yosemite Valley: (1) Allen, A. W. (Ed.) 31; (49) Matthes, F. E. 31

Geomagnetic tracing of buried river channels: (23) Ellsworth, E. W. 33a

Last Chance Gulch, Pliocene woods: (81) Webber, I. E. 33

Lava alterations, Lassen Volcanic National Park: (2) Anderson, C. A. 35

beds, National Monument: (72) Stearns, H. T. 28

block: (25) Finch, R. H. 33

cap, gold producer: (14) Chandler, J. W. 34

flow: (25) Finch, R. H. 31a
northern California: (25) Finch, R. H. 33a

tube, opal stalactites and stalagmites: (2) Anderson, C. A. 30a

Lassen Volcanic National Park: (2) Anderson, C. A. 34

Modoc Lava Bed quadrangle: (61) Powers, H. A. 32

radium content: (24) Evans, R. D. 35; (83) Williams, H. 35

Lead: (2) Allison, B. 35; (28) Gaylord, H. M. 34, 36; (33) Heikes, V. C. 32, 32a, 33; (37) Horton, F. W. 34; (52) Merrill, C. W. 34, 36; (52) Meyer, H. M. 32a, 33, 34; (54) Mineral Industry 33b; (58, 59) Pehrson, E. W. 29, 30, 33, 34a, 35a; (66) Santmyers, R. M. 31e; (86) Wright, C. W. 32

silver mining district, Darwin: (44) Knopf, A. 33b

Lepidocyclina: (18) Cushman, J. A. 20; (67) Schenck, H. G. 33a; 33b; (76) Taliaferro, N. L. 33b

Lewiston Dredge, property description: (63) Requa, L. K. 32

Lick Observatory station, earthquake registration: (3) Annis, W. 36, 36a; (11, 12) Byerly, P. 27, 27a, 28, 28a, 30, 30a, 31c, 32, 33d, 33e, 35, 35a, 35d, 36, 36a; (21) Dyk, R. 28, 30, 30a; (37) Hoskins, E. E. 36; (41) Jones, A. E. 27; (71) Sparks, N. R. 32, 33a, 33b; (83, 84) Wilson, J. T. 35b, 35d, 36

Lillis formation: (31) Hanna, G. D. 33
shale: (2) Anderson, F. M. 31; (15) Church, C. C. 31c

Lime: (8) Bowles, O. 35; (17) Coons, A. T. 32, 32f, 35; (32) Hatmaker, P. 32

Limestone: (8) Bowles, O. 34b
black-iron sulphide: (46) Laudermilk, J. D. 35; (85) Woodford, A. O. 35

brittle star, Miocene: (51) Merriam, C. W. 31

Calaveras Cement Company: (77) Townsand, R. H. 33

deposits, San Francisco region: (23) Eckel, E. C. 34

Eocene, orbitoid-bearing: (85) Woodring, W. P. 31a

McCloud, fauna and correlation: (82) Wheeler, H. E. 35a

mining at Crestmore: (64) Robotham, C. A. 34

Miocene: (51) Merriam, C. W. 31

rilled: (46) Laudermilk, J. D. 32, 32a; (85) Woodford, A. O. 32, 32a

weathering, San Francisco region: (42) Kelley, J. W. 33

Lincoln, gold placer mining: (39) Huttl, J. B. 35b

Lithium pegmatites, Pala: (21) Donnelly, M. 36

Lithophysae-bearing rhyolites, Santa Lucia range: (76) Taliaferro, N. L. 32; (78) Turner, R. E. 31

Llajas fauna, lower, Ventura County: (15) Clark, B. L. 34b
formation, Eocene: (51) McMasters, J. H. 32

Lode gold, prospecting: (28) Gardner, E. D. 35; (41) Johnson, F. W. 35

Lompoc beds, fossil fish: (30) Gudger, E. W. 33
oil field: (21) Dolman, S. G. 32
and gas storage: (35) Hodges, F. C. 32; (40) Johnson, A. M. 32

Lone Hill, Santa Clara County, geologic history: (64) Rogers, A. F. 31c

Mountain, landslides, mechanics: (17) Cogen, W. M. 36a

Long Beach earthquake: (11) Buwalda, J. P. 33; (16) Clements, T. 33; (20) Davis, W. M. 34; (26) Franz, S. I. 34; (29) Green, N. B. 33; (32) Heck, N. H. 33; (56) Nature 33; (57) Norris, A. 34; (62) Reeds, C. A. 33; (84, 85) Wood, H. O. 33, 33a, 33d; (23) Eaton, J. E. 33b

centers: (24) Eng. News-Rec. 33

cracks: (35) Hillis, D. 33

damage: (37) Horner, A. C. 33; (80) Wailes, C. D. 33

effect: (14) Chick, C. A. 33

longitudinal waves: (30) Gutenberg, B. 34

strong-motion records: (32) Heck, N. H. 33a

mollusca: (15) Clark, A. 33

oil fields: (18) Crown, W. J. 32; (34) Hight, W. 33; (38) Howard, P. J. 32; (60) Pierce, G. G. 32; (74) Stockman, L. P. 35

Repetto Hills: (62) Reed, R. D. 32

Long-beaked porpoise, Miocene: (42) Kellogg, R. 32

Long Valley Lake, Pleistocene: (50) Mayo, E. B. 34a

Longitudinal profiles of streams: (76) Taylor, E. 34; (85) Woodford, A. O. 33c

waves, Long Beach earthquake: (30) Gutenberg, B. 34

Los Alamos field: (14) Chappius, L. C. 31

Los Angeles Basin, excursion: (36) Hoots, H. W. 32

geology: (36) Hoots, H. W. 32b

late Quaternic: (23) Eckis, R. 35

oil companies' preparation for disaster: (30) Griffith, L. 29

development: (36) Hoots, H. W. 32c

fields: (3) Arnold, R. 27; (23) Emmons, W. H. 31; (47) Loel, W. 27; (74) Stockman, L. P. 35

geothermal variations: (13) Carlson, A. J. 31

Pliocene Cancer: (62) Rathbun, M. J. 32

conglomerates: (23) Edwards, E. C. 34

seismic reflection profile: (11) Buwalda, J. P. 35; (30) Gutenberg, B. 35

Oil—Cont.

(21) Dodd, H. V. 33; (42) Kaplow, E. J. 33

Lompoc: (21) Dolman, S. G. 32

Long Beach: (34) Hight, W. 33; (74) Stockman, L. P. 35; (18) Crown, W. J. 32; (37) Howard, P. J. 32; (60) Pierce, G. G. 32

Los Alamos: (14) Chappius, L. C. 31

Los Angeles Basin: (23) Emmons, W. H. 31; (3) Arnold, R. 27; (47) Loel, W. 27; (74) Stockman, L. P. 35

Lost Hills: (23) Emmons, W. H. 31

McKittrick: (23) Emmons, W. H. 31; (75) Taff, J. A. 33

water invasion: (40) Jensen, J. 30a, 31b; (72) Stevens, J. B. 30, 31

Manhattan Beach: (49) Marshall, W. C. 34

microscopic work: (62) Reed, R. D. 31b

Midway: (23) Emmons, W. H. 31

Sunset: (74) Stockman, L. P. 35

Montebello: (34) Hight, W. 33

Mount Poso: (21) Diepenbrock, A. 33; (34) Hight, W. 33; (74) Stockman, L. P. 35

Mountain View: (74) Stockman, L. P. 35

North Belridge: (23) Emmons, W. H. 31; (61) Preston, H. M. 32; (74) Stockman, L. P. 35

Operations: (10) Bush, R. D. 32

District No. 1: (38) Huguenin, E. 30

District No. 2: (76) Thoms, C. C. 31

District No. 3: (21) Dolman, S. G. 31a

District No. 4: (56) Musser, E. H. 31

District No. 5: (21) Dodd, H. V. 31a

palaeontologists textbook: (19) Davies, A. M. 35a

Petaluma: (23) Emmons, W. H. 31

Petrolia: (23) Emmons, W. H. 31

Playa del Rey: (5) Barton, C. L. 31; (41) Jones, P. H. 35; (36) Hoots, H. W. 35; (7) Blount, A. L. 35; (81) Westmith, J. N. 30b; (74) Stockman, L. P. 35; (34) Hight, W. 33

possibilities: (54) Millett, E. R. Jr. 35

production: (33) Herold, S. C. 33

Richfield: (34) Hight, W. 33; (74) Stockman, L. P. 35

Rincon: (74) Stockman, L. P. 35

Rosecrans: (34) Hight, W. 33

Round Mountain: (74) Stockman, L. P. 35

San Miguelto: (74) Stockman, L. P. 31, 35

Santa Barbara Mesa: (74) Stockman, L. P. 35

geology: (61) Powell, E. B. 34

Santa Clara: (23) Emmons, W. H. 31

Santa Fe Springs: (74) Stockman, L. P. 35

Santa Maria: (23) Emmons, W. H. 31

Sargent: (23) Emmons, W. H. 31

Seal Beach: (5) Barnes, R. M. 30; (74) Stockman, L. P. 35; (34) Hight, W. 33; (8) Bowes, G. H. 30

subsurface microscopic work: (62) Reed, R. D. 31b

Oil—Cont.

Summerland: (23) Emmons, W. H. 31

Sunset: (23) Emmons, W. H. 31

-Midway: (58) Pack, R. W. 20

Torrance: (34) Hight, W. 33; (74) Stockman, L. P. 35

unit development: (59) Petroleum Times, 32a

Venice: (59) Petroleum Times 30; (79) Van Couvering, M. 30

Ventura Avenue: (34) Hertel, F. W. 30, 31; (34) Hight, W. 33; (74) Stockman, L. P. 35

Ventura Basin: (74) Stockman, L. P. 35

waters: (45) Lahee, F. H. 34a (80) Washburn, C. W. 34; (40) Jensen, J. 34

Whittier fault: (57) Norris, B. B. 30

gain: (59) Petroleum World 35a

gas sands: (53) Miller, H. C. 32

genesis by radial axial pressure: (79) Uwatoko, K. 32

Huntington: (27) Gale, H. S. 34

industry: (59) Petroleum Times 34

development: (83) Wilhelm, V. H. 35

developments 1935: (83) Wilhelm, V. H. 36

Kern County: (59) Petroleum World 36b

marine drillings: (69) Siemon, J. H. 33; (28) Gilbert, J. C. 33

marine oil shale, Playa del Rey: (41) Jones, P. H. 35; (36) Hoots, H. W. 35; (7) Blount, A. L. 35

metamorphic rocks, San Gabriel Mountains: (9, 10) Brown, A. B. 32; (43) Kew, W. S. W. 32a

migration: (45) Lahee, F. H. 34; (63) Rich, J. L. 34

accumulation, origin: (63) Rich, J. L. 34

sandstone dykes through shales: (40) Jenkins, O. P. 31b

water encroachments at Casmalia: (60) Porter, W. W. 33

new fields outlook: (23) Eaton, J. E. 35

Northern California: (71) Stalder, W. 31, 31a, 32

occurrence, metamorphic rocks of San Gabriel Mountains: (9, 10) Brown, A. B. 32; (43) Kew, W. S. W. 32a

organic matter converted into: (77) Trask, P. D. 36

origin and accumulation: (15) Clark, F. R. 34; (63) Rich, J. L. 34

outlook for new fields: (23) Eaton, J. E. 35b

possibilities: (41) Johnston, J. H. 35

Rumsey Hills: (59) Petroleum World 32

Terra Bella, Tulare County: (75) Suverkrop, L. 31

problems, Kettleman Hills: (6) Bell, A. H. 31

production: (59) Pemberton, J. R. 35; (2) Allen, R. E. 35 (40) Jensen, J. 30b; (79) Uren, L. C. 36

Kettleman Hills: (54) Mills, B. 35a

San Joaquin Valley: (44) Kleinpell, W. D. 34

1929: (60) Petroleum World (London) 30e

1933: (59) Petroleum World 34b

San Onofre breccia: (85) Woodford, A. O. 26a
San Pedro, foraminifera: (55) Moyer, D. A. 29
  Pleistocene, Timms Point: (15) Clark, A. 31
  hills: (61) Putnam, W. C. 33; (47) Livingston, A. 33; (85) Woodring, W. P. 32d
  calcareous beds: (62) Reed, R. D. 31
  Pleistocene deposits: (43) Kew, W. S. W. 32b; (85) Woodring, W. P. 32c
  Pleistocene terraces: (85) Woodring, W. P. 35a
  Tertiary deposits: (43) Kew, W. S. W. 32b; (85) Woodring, W. P. 32c
San Rafael Mountains, limestones: (62) Reed, R. D. 35d
San Simeon quadrangle, geology: (76) Taliaferro, N. L. 35
Sanbornite: (9) Bradley, W. W. 32b; (51) Melhase, J. 35a
  Mariposa County: (64) Rogers, A. F. 32b
Sand: (38) Hughes, H. H. 32, 36; (1) Allan, M. 32; (18) Cornthwaite, M. A. 36a; (62) Reed, R. D. 30; (60) Phillips, E. R. 32
Sandblast, Sierra Nevada: (6) Blackwelder, E. 31
Sand-calcite crystals, Monterey County: (62) Reed, R. D. 25; (64) Rogers, A. F. 25
Sand concretions: (23) Edwards, S. C. 34
  Kettleman Hills: (9) Bramlette, M. N. 33a
  production: (25) Fisher, E. H. 31
  recent: (62) Reed, R. D. 30
Santa Ana Mountains: (61) Putnam, W. C. 33; (47) Livingston, A. 33
  foothills, unconformity exposed: (55) Moody, G. B. 35
  geology: (55) Moore, B. M. 30
  River Valley and Coastal Plain, rainfall; (76) Taylor, C. A. 30; (7) Blaney, H. F. 30; (86) Young, A. A. 30
Santa Barbara County, diatomite: (55) Mulryan, H. 36
  Elwood field, foraminifera: (70) Smith, W. M. 30
  Eocene marine algae, Sierra Blanca limestone: (38) Howe, M. A. 33
  Gaviota formation: (23) Effinger, W. L. 35
  geology, San Miguel Island: (9) Bremner, C. S. J. 33
  oil shale: (29) Gore, F. D. 27
  Pleistocene mollusca: (57) Oldroyd, I. S. 31; (29) Grant U. S. 31a
  Sierra Blanca limestone: (42) Keenan M. F. 32
  faulting: (35) Hill, M. L. 32
  Los Angeles, guide book: (43) Kew, W. S. W. 32
  Mesa oil field: (74) Stockman, L. P. 35
  geology: (61) Powell, E. B. 34
  Pliocene bird remains: (37) Howard, H. 31
  and Pleistocene mollusca: (29) Grant, U. S. 34; (75) Strong, A. M. 34a
  region, fauna: (25) Fisher, E. 30a
  water supply: (35) Hill, R. A. 32

Santa Catalina Island: (61) Randolph, G. C. 35
  marine terraces: (70) Smith, W. S. T. 33
  quarrying: (64) Roalfe, G. A. 32
Santa Clara County: (26) Franke, H. A. 30
  Lone Hill, geologic history: (64) Rogers, A. F. 31c
  oil field: (23) Emmons, W. H. 31
  River placers: (39) Jamison, C. E. 10
  Valley oil fields: (23) Emmons, W. H. 31
Santa Cruz Island, fauna: (25) Fisher, E. M. 30
  flora, Pleistocene: (14) Chaney, R. W. 34; (49) Mason, H. L. 34a
  geology: (9) Bremner, C. S. 32; (61) Rand, W. W. 31
  Pleistocene flora: (14) Chaney, R. W. 34; (49) Mason, H. L. 34a
Santa Fe Springs, edgewater encroachment: (81) Weaver, D. K. 31
  field, organic matter conversion into oil: (77) Trask, P. D. 36; (74) Stockman, L. P. 35
Santa Lucia Range, Lithophysae-bearing rhyolites: (76) Taliaferro, N. L. 32; (78) Turner, R. E. 31
  structure: (72) Stanton, W. L. 32
Santa Margarita conglomerate, Temblor Range: (62) Reed, R. D. 33c
Santa Maria district, lower Pliocene: (60) Porter, W. W. 32
  district, oil and structure: (17) Collom, R. E. 29
  oil field: (23) Emmons, W. H. 31
Santa Monica Bay, earthquake: (30) Gutenberg, B. 32a; (63) Richter, C. F. 32; (84) Wood, H. O. 32a
  sediments: (69) Shepard, F. P. 34; (50) McDonald, G. A. 34
  Mountains: (61) Putnam, W. C. 33; (47) Livingston, A. 33; (42) Kelley, V. C. 34
  eastern part, geology: (36) Hoots, H. W. 32a
  excursion: (36) Hoots, H. W. 32
  geology: (36) Hoots, H. W. 31a
  glacial epochs: (20) Davis, W. M. 32
  shore lines: (20) Davis, W. M. 31; (61) Putnam, W. C. 31; (63) Richters, G. L. 31; (20) Davis, W. M. 31a
  stratigraphy: (29) Grant, U. S. 34b; (71) Soper, E. K. 34
Santa Rita conglomerate, Temblor Range: (62) Reed, R. D. 33c
Santa Rosa Island, geology: (59) Petroleum World 32a; (55) Moody, G. B. 35a
  oil test: (59) Petroleum World 32a
Santa Susanna fauna, Ventura County: (15) Clark, B. L. 34b
Santa Ynez Range, Eocene foraminifera: (85) Woodring, W. P. 31a
  fossils: (56) Nelson, R. N. 31; (67) Schenck, H. G. 31a
  western, turritella variata zone: (85) Woodring, W. P. 31a
Saratoga chalk, foraminifera: (18) Cushman, J. A. 31b
Sargent oil field: (23) Emmons, W. H. 31
Scarp, Kern River: (6) Blackwelder, E. 29

Shales—Cont.

Miocene brown, Kettleman Hills: (67) Schenck, H. G. 31c
Monterey: (31) Hanna, G. D. 32
Reef Ridge, stratigraphy and foraminifera, Upper Miocene: (5) Barbat, W. F. 34a; (41) Johnson, F. L. 34
Sharks from Temblor group, Kern County: (41) Jordan, D. S. 31
tooth Hill: (79) Vander Eike, P. 36
diatoms: (31) Hanna, G. D. 32b
geology: (31) Hanna, G. D. 31b
Shasta County, Chico formation, fossil pearls: (65) Russell, R. D. 33
copper: (4) Averill, C. V. 33a
fusulinids: (82) Wheeler, H. E. 32
quadrangle: (3) Averill, C. V. 31
Sheep, mountain, Lyell Glacier: (5) Beatty, M. D. 33
Sheetfloods: (20) Davis, W. M. 33e
Shells, Eocene: (31) Hanna, G. D. 34, 36
Shorelines, Santa Monica Mountains: (20) Davis, W. M. 31a; (20) Davis, W. M. 31b; (61) Putnam, W. C. 31; (63) Richards, G. L. 31
Shoveltusker, Serbelodon burnhami: (57) Osborn, H. F. 33
Shrew, relic: (30) Grinnell, J. 32
Sienna: (66) Santmyers, R. M. 31d
Sierra Blanca limestone, Eocene: (42) Keenan, M. F. 32
marine algae, Santa Barbara County: (38) Howe, M. A. 33
foothills, Pliocene vertebrate fauna: (79) Vander Hoof, V. L. 33a
Tertiary, Pliohippus tantalus: (79) Vander Hoof, V. L. 33c
Nevada: (58) Panzer, W. 33
batholith: (16) Cloos, E. 32, 33, 33a, 33b
bed-rock complex: (76) Taliaferro, N. L. 32a, 32b
comagmatic region: (25) Fitch, A. A. 32a
east flank: (50) Mayo, E. B. 31
east slope: (7) Blackwelder, E. 33
fossil plants in auriferous gravels: (14) Chaney, R. W. 32
geologic report: (40) Jenkins, O. P. 32a
geology: (53) Miller, W. J. 31b
and geography: (49) Matthes, F. E. 33
geomorphology: (58) Panzer, W. 33
glacial and stream deposits: (7) Blackwelder, E. 32b
glaciation, Pleistocene: (7) Blackwelder, E. 31e
glaciers: (49) Matthes, F. E. 32
sand blast: (6) Blackwelder, E. 31
granitoid intrusives, Jurassic age: (35) Hinds, N. E. A. 34b
intrusions and their wall rocks: (50) Mayo, E. B. 35
structure: (4) Balk, R. 31; (16) Cloos, E. 31
Milton formation: (16) Clark, S. G. 33
mountaineering: (43) King, C. 35
peneplain: (58) Panzer, W. 33
Pleistocene glaciation: (6) Blackwelder, E. 31a
Pluton: (16) Cloos, E. 31b, 33c
resurrection of early surfaces: (40) Jenkins, O. P. 34

Sierra Blanca—Cont.

scarps: (31) Hake, B. F. 32
southern part, eastern slope geology: (44) Knopf, A. 27
western slope: (49) Matthes, F. E. 33a
Signal Hill: (39) Jay, M. 31; (65) Russell, S. 31
Silica: (60) Phillips, E. R. 32a; (66) Santmyers, R. M. 31a
deposits: (65, 66) Sampson, R. J. 31; (77) Tucker, W. B. 31a
-fluorite pseudomorphs: (55) Murdoch, J. 36
Siliceous shale formation, southern California: (62) Reed, R. D. 31a
Silicoflagellates, Kreyenhagen shale: (31) Hanna, G. D. 31
Silver: (9) Bratter, H. M. 36; (22) Dunlop, J. P. 32, 36; (33) Henderson, C. W. 36a; (52) Merrill, C. W. 36; (28) Gaylord, H. M. 36, 34; (47) Lilley, E. R. 36; (37) Horton, F. W. 34; (33) Heikes, V. C. 32, 32a, 33; (52) Merrill, C. W. 34; (33) Henderson, C. W. 32; (46) Lawrie, H. N. 30
-lead mining district, Darwin: (44) Knopf, A. 33b
yield, copper ores: (58) Pehrson, E. W. 34
Simi Valley, carnivora, Sespe Eocene: (73) Stock, C. 33
Eocene stratigraphy: (77) Tolman, F. B. 34; (72, 73) Stipp, T. F. 34
vertebrate fauna: (74) Stock, C. 34a
Sespe Eocene mammals: (73) Stock, C. 32d
vertebrate fauna: (74) Stock, C. 34a
Tertiary deposits: (85) Woodring, W. P. 36
Siskiyou County, chromite: (64) Rogers, A. F. 31
Cretaceous deposits: (2) Anderson, F. M. 31a
geology: (50) Maxson, J. H. 33a
mines, mineral resources: (4) Averill, C. V. 35
Siskiyous, south: (38) Huttl, J. B. 34
Slate: (47) Lilley, E. R. 36; (8) Bowles, O. 32a, 34b; (17) Coons, A. T. 32c
Hot Springs, Cretaceous beds: (57) Nomland, J. O. 32; (67) Schenck, H. G. 32
Soapstone: (8) Bowles, O. 31; (23) Emery, A. H. 33; (75) Stoddard, B. H. 32c, 33; (8) Bowles, O. 32d, 31b, 34b; (38) Hughes, H. H. 32a; (54) Mineral Industry 34a
ground: (23) Emery, A. H. 33; (74) Stoddard, B. H. 32
Soda, alkaline lake brines supply: (35) Hirschkind, W. 31
Sodium compounds: (47) Lilley, E. R. 36
natural: (17) Coons, A. T. 32
salts: (48) Manning, P. D. V. 31
sulphate: (78) Tyler, P. M. 35b
Soil erosion: (81) Weir, W. W. 32
hillocks, erosional: (51) Melton, F. A. 34
investigation, southern California: (48) Lowry, H. 31
rating: (75) Storie, R. E. 36; (81) Weir, W. W. 36
survey, distribution: (2) Allyne, A. B. 32

PUBLICATIONS OF THE DIVISION OF MINES

During the past fifty-six years, in carrying out the provisions of the organic act creating the former California State Mining Bureau, there have been published many reports, bulletins and maps which go to make up a library of detailed information on the mineral industry of the State, a large part of which could not be duplicated from any other source.

One feature that has added to the popularity of the publications is that many of them have been distributed without cost to the public, and even the more elaborate ones have been sold at a price which barely covers the cost of printing.

Owing to the fact that funds for the advancing of the work of this department have usually been limited, the reports and bulletins mentioned are printed in limited editions many of which are now entirely exhausted.

Copies of such publications are available for reference, however, in the offices of the Division of Mines, in the Ferry Building, San Francisco; State Building, Los Angeles; State Office Building, Sacramento; Redding; and Division of Oil and Gas at Santa Barbara, Taft, Bakersfield, Coalinga. They may also be found in many public, private and technical libraries in California and other states and foreign countries.

A catalog of all publications from 1880 to 1917, giving a synopsis of their contents, is issued as Bulletin No. 77.

Publications in stock may be obtained postpaid by addressing any of the above offices and enclosing the requisite amount in the case of publications that have a list price. Only coin, stamps or money orders should be sent, and it will be appreciated if remittance is made in this manner rather than by personal check.

Money orders should be made payable to the Division of Mines.

NOTE.—The Division of Mines frequently receives requests for some of the early Reports and Bulletins now out of print, and it will be appreciated if parties having such publications and wishing to dispose of them will advise this office.

REPORTS

Asterisks (**) indicate the publication is out of print.

Price
Postpaid

**First Annual Report of the State Mineralogist, 1880, 43 pp. Henry
G. Hanks _____   ____
**Second Annual Report of the State Mineralogist, 1882, 514 pp., 4 illustra-
tions, 1 map. Henry G. Hanks_____   ____
**Third Annual Report of the State Mineralogist, 1883, 111 pp., 21 illustra-
tions. Henry G. Hanks_____   ____
**Fourth Annual Report of the State Mineralogist, 1884, 410 pp., 7 illustra-
tions. Henry G. Hanks_____   ____
**Fifth Annual Report of the State Mineralogist, 1885, 234 pp., 15 illustra-
tions, 1 geological map. Henry G. Hanks_____   ____
Sixth Annual Report of the State Mineralogist, Part I, 1886, 145 pp., 3
illustrations, 1 map. Henry G. Hanks_____ $0.70
Part II, 1887, 222 pp., 36 illustrations. William Irelan, Jr._____   .70
**Seventh Annual Report of the State Mineralogist, 1887, 315 pp. William
Irelan, Jr. _____   ____
**Eighth Annual Report of the State Mineralogist, 1888, 948 pp., 122 illus-
trations. William Irelan, Jr._____   ____
Ninth Annual Report of the State Mineralogist, 1889, 352 pp., 57 illustra-
tions, 2 maps. William Irelan, Jr._____   1.15
**Tenth Annual Report of the State Mineralogist, 1890, 983 pp., 179 ilus-
trations, 10 maps. William Irelan, Jr._____   ____
Eleventh Report (First Biennial) of the State Mineralogist, for the two
years ending September 15, 1892, 612 pp., 73 illustrations, 4 maps
William Irelan, Jr._____   1.25
**Twelfth Report (Second Biennial) of the State Mineralogist, for the two
years ending September 15, 1894, 541 pp., 101 illustrations, 5 maps.
J. J. Crawford _____   ____
**Thirteenth Report (Third Biennial) of the State Mineralogist, for the
two years ending September 15, 1896, 726 pp., 93 illustrations, 1
map. J. J. Crawford_____   ____
Chapters of the State Mineralogist's Report, Biennial Period, 1913-1914,
Fletcher Hamilton:
Mines and Mineral Resources, Amador, Calaveras and Tuolumne Counties,
172 pp., paper _____   .60
Mines and Mineral Resources, Colusa, Glenn, Lake, Marin, Napa, Solano,
Sonoma and Yolo Counties, 208 pp., paper_____   .60
Mines and Mineral Resources, Del Norte, Humboldt and Mendocino Coun-
ties, 59 pp., paper _____   .35
**Mines and Mineral Resources, Fresno, Kern, Kings, Madera, Mariposa,
Merced, San Joaquin and Stanislaus Counties, 220 pp., paper_____   ____
Mines and Mineral Resources of Imperial and San Diego Counties, 113
pp., paper _____   .50
Mines and Mineral Resources, Shasta, Siskiyou and Trinity Counties,
180 pp., paper _____   .60
Fourteenth Report of the State Mineralogist, for the Biennial Period 1913-
1914, Fletcher Hamilton, 1915:
A General report on the Mines and Mineral Resources of Amador,
Calaveras, Tuolumne, Colusa, Glenn, Lake, Marin, Napa, Solano,
Sonoma, Yolo, Del Norte, Humboldt, Mendocino, Fresno, Kern,
Kings, Madera, Mariposa, Merced, San Joaquin, Stanislaus, San
Diego, Imperial, Shasta, Siskiyou and Trinity Counties, 974 pp., 275
illustrations, cloth _____   3.00
Chapters of the State Mineralogist's Report, Biennial Period, 1915-1916.
Fletcher Hamilton:
Mines and Mineral Resources, Alpine, Inyo and Mono Counties, 176 pp.,
paper _____   .75
Mines and Mineral Resources, Butte, Lassen, Modoc, Sutter and Tehama
Counties, 91 pp., paper_____   .55
Mines and Mineral Resources, El Dorado, Placer, Sacramento and Yuba
Counties, 198 pp., paper_____   .75
Mines and Mineral Resources, Monterey, San Benito, San Luis Obispo,
Santa Barbara and Ventura Counties, 183 pp., paper_____   .75

REPORTS—Continued

Asterisks (**) indicate the publication is out of print.

Price
Postpaid

Mines and Mineral Resources, Los Angeles, Orange and Riverside Counties, 136 pp., paper_____ $0.60
Mines and Mineral Resources, San Bernardino and Tulare Counties, 186 pp., paper _____ .75
**Fifteenth Report of the State Mineralogist, for the Biennial Period 1915-1916, Fletcher Hamilton, 1917:
A General Report on the Mines and Mineral Resources of Alpine, Inyo, Mono, Butte, Lassen, Modoc, Sutter, Tehama, Placer, Sacramento, Yuba, Los Angeles, Orange, Riverside, San Benito, San Luis Obispo, Santa Barbara, Ventura, San Bernardino and Tulare Counties, 990 pp., 413 illustrations, cloth_____ ____
Chapters of the State Mineralogist's Report, Biennial Period, 1917-1918, Fletcher Hamilton:
Mines and Mineral Resources of Nevada County, 270 pp., paper_____ .90
Mines and Mineral Resources of Plumas County, 188 pp., paper_____ .60
Mines and Mineral Resources of Sierra County, 144 pp., paper_____ .60
Seventeenth Report of the State Mineralogist, 1920, 'Mining in California during 1920,' Fletcher Hamilton; 562 pp., 71 illustrations, cloth_____ 2.00
Eighteenth Report of the State Mineralogist, 1922, 'Mining in California,' Fletcher Hamilton. Chapters published monthly beginning with January, 1922:
**January, **February, March, April, **May, June, July, August, September, October, November, December, 1922_____ .30
Chapters of Nineteenth Report of the State Mineralogist, 'Mining in California,' Fletcher Hamilton and Lloyd L. Root. January, February, March, September, 1923_____ .30
Chapters of Twentieth Report of the State Mineralogist, 'Mining in California,' Lloyd L. Root. Published quarterly. January, April, July, October, 1924, per copy_____ .30
Chapters of Twenty-first Report of the State Mineralogist, 'Mining in California,' Lloyd L. Root. Published quarterly:
January, 1925, Mines and Mineral Resources of Sacramento, Monterey and Orange Counties_____ .30
April, 1925, Mines and Mineral Resources of Calaveras, Merced, San Joaquin, Stanislaus and Ventura Counties_____ .30
July, 1925, Mines and Mineral Resources of Del Norte, Humboldt and San Diego Counties _____ .30
October, 1925, Mines and Mineral Resources of Siskiyou, San Luis Obispo and Santa Barbara Counties_____ .30
Chapters of Twenty-second Report of the State Mineralogist, 'Mining in California, Lloyd L. Root. Published quarterly:
January, 1926, Mines and Mineral Resources of Trinity and Santa Cruz Counties _____ .30
April, 1926, Mines and Mineral Resources of Shasta, San Benito and Imperial Counties _____ .35
July, 1926, Mines and Mineral Resources of Marin and Sonoma Counties .30
**October, 1926, Mines and Mineral Resources of El Dorado and Inyo Counties, also report on Minaret District, Madera County_____ ____
Chapters of Twenty-third Report of the State Mineralogist, 'Mining in California,' Lloyd L. Root. Published quarterly:
January, 1927, Mines and Mineral Resources of Contra Costa County; Santa Catalina Island_____ .35
April, 1927, Mines and Mineral Resources of Amador and Solano Counties .30
**July, 1927, Mines and Mineral Resources of Placer and Los Angeles Counties _____ ____
October, 1927, Mines and Mineral Resources of Mono County_____ .30
Chapters of Twenty-fourth Report of the State Mineralogist, 'Mining in California,' Lloyd L. Root. Published quarterly:
January, 1928, Mines and Mineral Resources of Tuolumne County_____ .30
April, 1928, Mines and Mineral Resources of Mariposa County_____ .30
July, 1928, Mines and Mineral Resources of Butte and Tehama Counties .30

REPORTS—Continued

Asterisks (**) indicate the publication is out of print.

REPORTS—Continued

Asterisks (**) indicate the publication is out of print.

Price
Postpaid

Chapters of Report XXIX, 1933 (quarterly: titled 'California Journal of Mines and Geology,' containing the following:

January-April. Gold Deposits of the Redding and Weaverville Quadrangles. Geologic Formations of the Redding-Weaverville District, Northern California. Geology of Portions of Del Norte and Siskiyou Counties. Applications of Geology to Civil Engineering. The Lakes of California. Discovery of Piedmontite in the Sierra Nevada. Tracing 'Buried River' Channel Deposits by Geomagnetic Methods. Geologic Map of Redding-Weaverville District, showing gold mines and prospects. Geologic Map showing various mines and prospects of part of Del Norte and Siskiyou Counties_____ $0.90

July-October. Gold Resources of Kern County. Limestone Deposits of the San Francisco Region. Limestone Weathering and Plant Associations of the San Francisco Region. Booming. Death Valley National Monument, California. Placer Mining Districts, Senate Bill 480. Navigable Waters, Assembly Bill 1543_____ .90

Chapters of Report XXX, 1934 (quarterly): titled 'California Journal of Mines and Geology,' containing the following:

January. Resurrection of Early Surfaces in the Sierra Nevada. Geology and Mineral Resources of Northeastern Madera County. Geology and Mineral Deposits of Laurel and Convict Basins, Southwestern Mono County. Notes on Sampling as Applied to Gold Quartz Deposits_____ .50

April-July. Elementary Placer Mining in California and Notes on the Milling of Gold Ores_____ .90

October. Current Mining Developments in Northern California. Current Mining Activity in Southern California. Geology and Mineral Resources of the Julian District, San Diego County. Geology and Mineral Resources of Elizabeth Lake Quadrangle. Dry Placers of Northern Mojave Desert. Biennial Report of State Mineralogist. Assessment Work Within Withdrawn Areas_____ .50

Chapters of Report XXI, 1935 (quarterly): titled 'California Journal of Mines and Geology,' containing the following:

January. Review of Gold Mining in East-Central, 1934. Current Mining Activities in the San Francisco District with Special Reference to Gold. Geological Investigation of the Clays of Riverside and Orange Counties, Southern California. Information regarding Mining Loans by the Reconstruction Finance Corporation_____ .50

April. A Geologic Section Across the Southern Peninsular Range of California. New Technique Applicable to the Study of Placers. Grubstake Permits _____ .50

July. Mines and Mineral Resources of Siskiyou County (with map). Dams for Hydraulic Mining Debris. Leasing System as Applied to Metal Mining. Mine Financing in California. New Laws Make Radical Change in Mining Rights_____ .50

October. Mines and Mineral Resources of San Luis Obispo County. Mineral Resources of Portions of Monterey and Kings Counties. Mining Activity at Soledad Mountain and Middle Buttes—Mojave District, Kern County. Geology of a Portion of the Perris Block, Southern California. Mineral Resources of a Portion of the Perris Block, Riverside County _____ .50

Chapters of Report XXXII, 1936 (quarterly): titled 'California Journal of Mines and Geology,' containing the following:

January. Gold Mines of Placer County, including Drag-line Dredges. Geologic Report on Borax Lake, California_____ .50

April. Geology, Mining and Processing of Diatomite at Lompoc, Santa Barbara County. Essentials in Developing and Financing a Prospect into a Mine. Gold-bearing Veins of Meadow Lake District, Nevada County. Semi-Precious Gem Stone Collection in Division Museum__ .50

July. Mines and Mineral Resources of Calaveras County. Mining in California by Power Shovel. Assessment Work on Mining Claims Within Withdrawn Areas. Joshua Tree National Monument. Cost

REPORTS—Continued

Asterisks (**) indicate the publication is out of print.

Price
Postpaid

of Producing Quicksilver at a California Mine in 1931-1932. The
Age of Mineral Utilization _____ $0.50

October. Mineral Resources of Lassen and Modoc Counties. Mechanics of
Lone Mountain Landslides, San Francisco. Biennial Report of the
State Mineralogist, Properties and Industrial Applications of Opaline
Silica _____ .50

Chapters of Report XXXIII, 1937 (quarterly) : titled 'California Journal
of Mines and Geology,' containing the following :

January. Source Data of the Geologic Map of California, January, 1937.
The Geology of Quicksilver Ore Deposits. Prospecting for Lode
Gold _____ .50

April. Mineral Resources of Plumas County (with Geologic Map).
List of preferred mineral names. New Placer Mining Debris Law .50

July. Mineral Resources of Los Angeles County (with map showing
principal Mines and Oil Fields.) Geology and mineral deposits
of the Western San Gabriel Mountains, Los Angeles County_____ .50

Subscription, $2.00 postpaid in advance (by calendar year only).

Chapters of State Oil and Gas Supervisor's Report :

Summary of Operations—California Oil Fields, July, 1918, to March,
1919 (one volume) _____ Free

Summary of Operations—California Oil Fields. Published monthly,
beginning April, 1919 :

**April, **May, **June, **July, **August, **September, **October,
**November, **December, 1919_____ ____

**January, **February, **March, **April, **May, **June, **July,
**August, **September, **October, **November, **December, 1920_ ____

January, **February, **March, April, **May, **June, **July, August,
**September, **October, **November, **December, 1921_____ Free

January, February, March, April, May, June, **July, **August, September, **October, **November, December, 1922_____ Free

January, February, **March, **April, May, **June, **July, August,
September, **October, November, **December, 1923_____ Free

January, February, March, April, May, June, **July, August, September, October, November, December, 1924_____ Free

January, February, March, April, May, June, July, August, September,
October, November, December, 1925_____ Free

January, February, March, April, May, June, July, August, September,
October, November, December, 1926_____ Free

January, February, March, April, May, June, July, August, September,
October, November, December, 1927_____ Free

January, February, March, April, **May, June, July, August, September, October, **November, **December, 1928_____ Free

January, February, March, April, May, June, July-August-September,
October-November-December, 1929 _____ Free
(Published quarterly beginning July, 1929)

January-February-March, April-May-June, July-August-September, October-November-December, 1930_____ Free

January-February-March, April-May-June, July-August-September, 1931 Free

January, February, March, April, May, June, July, August, September,
October, November, December, 1932_____ Free

January, February, March, 1933_____ Free

April, May, June, 1933_____ Free

July, August, September, 1933_____ Free

October-November-December, 1933 _____ Free

January-February-March, 1934_____ Free

April-May-June, 1934 _____ Free

July-August-September, 1934 _____ Free

October-November-December, 1934 _____ Free

January-February-March, 1935 _____ Free

April-May-June, 1935 _____ Free

BULLETINS

Asterisks (**) indicate the publication is out of print.

Price
Postpaid

**Bulletin No. 1. Description of Some Desiccated Human Remains, by Winslow Anderson. 1888, 41 pp., 6 illustrations_____ ____

**Bulletin No. 2. Methods of Mine Timbering, by W. H. Storms. 1894, 58 pp., 75 illustrations_____ ____

**Bulletin No. 3. Gas and Petroleum Yielding Formations of Central Valley of California, by W. L. Watts. 1894, 100 pp., 13 illustrations, 4 maps ____

Bulletin No. 4. Catalogue of California Fossils, by J. G. Cooper, 1894, 73 pp., 67 illustrations. (Part I was published in the Seventh Annual Report of the State Mineralogist, 1887)_____ $0.10

**Bulletin No. 5. The Cyanide Process, 1894, by Dr. A. Scheidel. 140 pp., 46 illustrations _____ ____

**Bulletin No. 6. California Gold Mill Practices, 1895, by E. B. Preston, 85 pp., 46 illustrations_____ ____

**Bulletin No. 7. Mineral Production of California, by Counties, for the year 1894, by Charles G. Yale. Tabulated sheet_____ ____

**Bulletin No. 8. Mineral Production of California, by Counties, for the year 1895, by Charles G. Yale. Tabulated sheet_____ ____

Bulletin No. 9. Mine Drainage, Pumps, etc., by Hans C. Behr. 1896, 210 pp., 206 illustrations_____ .75

Bulletin No. 10. A Bibliography Relating to the Geology, Paleontology and Mineral Resources of California, by Anthony W. Vogdes. 1896, 121 pp. _____ .50

**Bulletin No. 11. Oil and Gas Yielding Formations of Los Angeles, Ventura and Santa Barbara Counties, by W. L. Watts. 1897, 94 pp., 6 maps, 31 illustrations_____ ____

Bulletin No. 12. Mineral Production of California, by Counties, for 1896, by Charles G. Yale. Tabulated sheet_____ .10

**Bulletin No. 13. Mineral Production of California, by Counties, for 1897, by Charles G. Yale. Tabulated sheet_____ ____

**Bulletin No. 14. Mineral Production of California, by Counties, for 1898, by Charles G. Yale _____ ____

**Bulletin No. 15. Map of Oil City Fields, Fresno County, by John H. Means, 1899 _____ ____

**Bulletin No. 16. The Genesis of Petroleum and Asphaltum in California, by A. S. Cooper. 1899, 39 pp., 29 illustrations_____ ____

**Bulletin No. 17. Mineral Production of California, by Counties, for 1899, by Charles G. Yale. Tabulated sheet_____ ____

**Bulletin No. 18. Mother Lode Region of California, by W. H. Storms, 1900, 154 pp., 49 illustrations_____ ____

**Bulletin No. 19. Oil and Gas Yielding Formations of California, by W. L. Watts. 1900, 236 pp., 60 illustrations, 8 maps_____ ____

**Bulletin No. 20. Synopsis of General Report of State Mining Bureau, by W. L. Watts. 1901, 21 pp. This bulletin contains a brief statement of the progress of the mineral industry in California for the four years ending December, 1899_____ ____

Bulletin No. 21. Mineral Production of California by Counties, by Charles G. Yale. 1900. Tabulated sheet_____ .10

Bulletin No. 22. Mineral Production of California for Fourteen Years, by Charles G. Yale. 1900. Tabulated sheet_____ .10

Bulletin No. 23. The Copper Resources of California, by P. C. DuBois, F. M. Anderson, J. H. Tibbits and G. A. Tweedy. 1902, 282 pp., 69 illustrations, 9 maps _____ .75

**Bulletin No. 24. The Saline Deposits of California, by G. E. Bailey. 1902, 216 pp., 99 illustrations, 5 maps_____ ____

Bulletin No. 25. Mineral Production of California, by Counties, for 1901, by Charles G. Yale. Tabulated sheet_____ .10

Bulletin No. 26. Mineral Production of California for the Past Fifteen Years, by Charles G. Yale. 1902. Tabulated sheet_____ .10

**Bulletin No. 27. The Quicksilver Resources of California, by William Forstner. 1903, 273 pp., 144 illustrations, 8 maps_____ ____

Bulletin No. 28. Mineral Production of California for 1902, by Charles G. Yale. Tabulated sheet _____ .10

BULLETINS—Continued

Asterisks (**) indicate the publication is out of print.

Price
Postpaid

Bulletin No. 29. Mineral Production of California for Sixteen years,
by Charles G. Yale. 1903. Tabulated sheet_____ $0.10

**Bulletin No. 30. Bibliography Relating to the Geology, Paleontology and
Mineral Resources of California, by A. W. Vogdes. 1903, 290 pp.___ ____

**Bulletin No. 31. Chemical Analyses of California Petroleum, by H. N.
Cooper. 1904. Tabulated sheet_____ ____

**Bulletin No. 32. Production and Use of Petroleum in California, by Paul
W. Prutzman. 1904, 230 pp., 116 illustrations, 14 maps_____ ____

**Bulletin No. 33. Mineral Production of California, by Counties, for 1903,
by Charles G. Yale. Tabulated sheet_____ ____

**Bulletin No. 34. Mineral Production of California for Seventeen Years,
by Charles G. Yale. 1904. Tabulated sheet_____ ____

**Bulletin No. 35. Mines and Minerals of California, by Charles G. Yale.
1904, 55 pp., 20 county maps. Relief map of California_____ ____

**Bulletin No. 36. Gold Dredging in California, by J. E. Doolittle. 1905.
120 pp., 66 illustrations, 3 maps_____ ____

**Bulletin No. 37. Gems, Jewelers' Materials, and Ornamental Stones of
California, by George F. Kunz. 1905, 168 pp., 54 illustrations_____ ____

**Bulletin No. 38. Structural and Industrial Materials of California, by
Wm. Forstner, T. C. Hopkins, C. Naramore and L. H. Eddy. 1906,
412 pp., 150 illustrations, 1 map_____ ____

Bulletin No. 39. Mineral Production of California, by Counties, for 1904,
by Charles G. Yale. Tabulated sheet_____ .10

Bulletin No. 40. Mineral Production of California for Eighteen Years,
by Charles G. Yale. 1905. Tabulated sheet_____ .10

Bulletin No. 41. Mines and Minerals of California for 1904, by Charles
G. Yale. 1905, 54 pp., 20 county maps_____ .10

Bulletin No. 42. Mineral Production of California, by Counties, 1905,
by Charles G. Yale. Tabulated sheet_____ .10

Bulletin No. 43. Mineral Production of California for Nineteen Years,
by Charles G. Yale. Tabulated sheet_____ .10

Bulletin No. 44. California Mines and Minerals for 1905, by Charles
G. Yale. 1907, 31 pp., 20 county maps_____ .10

**Bulletin No. 45. Auriferous Black Sands of California, by J. A. Edman.
1907. 10 pp. _____ ____

**Bulletin No. 46. General Index of Publications of the California State
Mining Bureau, by Charles G. Yale. 1907, 54 pp._____ ____

**Bulletin No. 47. Mineral Production of California, by Counties, 1906,
by Charles G. Yale. Tabulated sheet_____ ____

**Bulletin No. 48. Mineral Production of California for Twenty Years, by
Charles G. Yale. 1906_____ ____

**Bulletin No. 49. Mines and Minerals of California for 1906, by Charles
G. Yale. 34 pp._____ ____

Bulletin No. 50. The Copper Resources of California, 1908, by A. Haus-
mann, J. Kruttschnitt, Jr., W. E. Thorn and J. A. Edman, 366 pp.,
74 illustrations. (Revised edition)_____ 1.25

Bulletin No. 51. Mineral Production of California, by Counties, 1907,
by D. H. Walker. Tabulated sheet_____ .10

Bulletin No. 52. Mineral Production of California for Twenty-one Years,
by D. H. Walker, 1907. Tabulated sheet_____ .10

Bulletin No. 53. Mineral Production of California for 1907, with County
Maps, by D. H. Walker. 62 pp _____ .10

**Bulletin No. 54. Mineral Production of California, by Counties, by D. H.
Walker, 1908. Tabulated sheet_____ ____

Bulletin No. 55. Mineral Production of California for Twenty-two Years,
by D. H. Walker, 1908. Tabulated sheet_____ .10

**Bulletin No. 56. Mineral Production for 1908, with County Maps and
Mining Laws of California, by D. H. Walker, 78 pp._____ ____

**Bulletin No. 57. Gold Dredging in California, by W. B. Winston and
Chas. Janin. 1910, 312 pp., 239 illustrations, 10 maps_____ ____

Bulletin No. 58. Mineral Production of California, by Counties, by D.
H. Walker. 1909. Tabulated sheet_____ .10

BULLETINS—Continued

Asterisks (**) indicate the publication is out of print.

Price
Postpaid

Bulletin No. 59. Mineral Production of California for Twenty-three
Years, by D. H. Walker. 1909. Tabulated sheet_____ **$0.10**
**Bulletin No. 60. Mineral Production for 1909, with County Maps and
Mining Laws of California, by D. H. Walker. 94 pp._____ ----
Bulletin No. 61. Mineral Production of California, by Counties, for 1910,
by D. H. Walker. Tabulated sheet_____ .10
**Bulletin No. 62. Mineral Production of California for Twenty-four
Years, by D. H. Walker. 1910. Tabulated sheet_____ ----
**Bulletin No. 63. Petroleum in Southern California, by P. W. Prutzman.
1912, 430 pp., 41 illustrations, 6 maps_____ ----
Bulletin No. 64. Mineral Production for 1911, by E. S. Boalich. 49 pp. .15
Bulletin No. 65. Mineral Production for 1912, by E. S. Boalich. 64 pp. .25
**Bulletin No. 66. Mining Laws of the United States and California. 1914,
89 pp. _____ ----
**Bulletin No. 67. Minerals of California, by Arthur S. Eakle. 1914, 226
pp _____ ----
Bulletin No. 68. Mineral Production for 1913, with County Maps and
Mining Laws, by E. S. Boalich. 160 pp._____ .25
**Bulletin No. 69. Petroleum Industry of California, with Folio of Maps
(18 by 22), by R. P. McLaughlin and C. A. Waring. 1914, 519 pp.,
13 illustrations, 83 figs. [18 plates in accompanying folio.]_____ ----
Bulletin No. 70. Mineral Production for 1914, with County Maps and
Mining Laws. 184 pp. _____ .25
Bulletin No. 71. Mineral Production for 1915, with County Maps and
Mining Laws, by Walter W. Bradley, 193 pp. 4 illustrations_____ .25
**Bulletin No. 72. The Geologic Formations of California, by James Perrin
Smith. 1916, 47 pp._____ ----
**Reconnaissance Geologic Map (of which Bulletin 72 is explanatory), in
23 colors. Scale: 1 inch = 12 miles. Mounted_____ ----
**Bulletin No. 73. First Annual Report of the State Oil and Gas Super-
visor of California for the fiscal year 1915-16, by R. P. McLaughlin.
278 pp., 26 illustrations_____ ----
Bulletin No. 74. Mineral Production of California in 1916, with County
Maps, by Walter W. Bradley. 179 pp., 12 illustrations_____ .25
**Bulletin No. 75. United States and California Mining Laws. 1917, 115
pp., paper _____ ----
Bulletin No. 76. Manganese and Chromium in California, by Walter W.
Bradley, Emile Huguenin, C. A. Logan, W. B. Tucker and C. A. War-
ing. 1918, 248 pp., 51 illustrations, 5 maps, paper_____ .60
Bulletin No. 77. Catalogue of Publications of California State Mining
Bureau, 1880-1917, by E. S. Boalich. 44 pp., paper_____ Free
Bulletin No. 78. Quicksilver Resources of California, with a Section on
Metallurgy and Ore-Dressing, by Walter W. Bradley. 1919, 389
pp., 77 photographs and 42 plates (colored and line cuts), cloth____ 1.90
Bulletin No. 79. Magnesite in California, by Walter W. Bradley. 1925,
147 pp., 62 photographs, 11 line cuts and maps, cloth_____ 1.10
†Bulletin No. 80. Tungsten. Molybdenum and Vanadium in California.
(In preparation.)
†Bulletin No. 81. Foothill Copper Belt of California. (In preparation.)
**Bulletin No. 82. Second Annual Report of the State Oil and Gas Super-
visor, for the Fiscal Year 1916-1917, by R. P. McLaughlin. 1918,
412 pp., 31 illustrations, cloth_____ ----
Bulletin No. 83. California Mineral Production for 1917, with County
Maps, by Walter W. Bradley. 179 pp., paper_____ .15
**Bulletin No. 84. Third Annual Report of the State Oil and Gas Super-
visor, for the Fiscal Year 1917-1918, by R. P. McLaughlin. 1918, 617
pp., 28 illustrations, cloth_____ ----
**Bulletin No. 85. Platinum and Allied Metals in California, by C. A.
Logan, 1919. 10 photographs, 4 plates, 120 pp., paper_____ ----

_____

† Not yet published.

BULLETINS—Continued

Asterisks (**) indicate the publication is out of print.

Price
Postpaid

Bulletin No. 86.   California Mineral Production for 1918, with County
Maps, by Walter W. Bradley.   1919, 212 pp., paper_____   Free

Bulletin No. 87.   Commercial Minerals of California, with notes on their
uses, distribution, properties, ores, field tests, and preparation for
market, by W. O. Castello.   1920, 124 pp., paper_____ $0.75

Bulletin No. 88.   California Mineral Production for 1919, with County
Maps, by Walter W. Bradley.   1920, 204 pp., paper_____   Free

**Bulletin No. 89.   Petroleum Resources of California, with Special Ref-
erence to Unproved Areas, by Lawrence Vander Leck.   1921, 12
figures, 6 photographs, 6 maps in pocket, 186 pp., cloth _____   ____

Bulletin No. 90.   California Mineral Production for 1920, with County
Maps, by Walter W. Bradley.   1921, 218 pp., paper _____   .25

**Bulletin No. 91.   Minerals of California, by Arthur S. Eakle.   1923, 328
pp., cloth _____   ____

**Bulletin No. 92.   Gold Placers of California, by Charles S. Haley.   1923,
167 pp., 36 photographs and 7 plates (colored and line cuts, also
geological map), cloth _____   ____

Bulletin No. 93.   California Mineral Production for 1922, by Walter W.
Bradley.   1923, 188 pp., paper_____   .15

Bulletin No. 94.   California Mineral Production for 1923, by Walter W.
Bradley.   1924, 162 pp., paper_____   .25

Bulletin No. 95.   Geology and Ore Deposits of the Randsburg Quadrangle,
by Carlton D. Hulin.   1925, 152 pp., 49 photographs, 13 line cuts,
1 colored geologic map, cloth _____   2.75

Bulletin No. 96.   California Mineral Production for 1924, by Walter W.
Bradley.   1925, 173 pp., paper_____   .15

**Bulletin No. 97.   California Mineral Production for 1925, by Walter W.
Bradley.   1926, 172 pp., paper_____   ____

Bulletin No. 98.   American Mining Law, by A. H. Ricketts, 1931, 811
pp., flexible leather _____   3.25

Bulletin No. 99.   Clay Resources and Ceramic Industry of California, by
Waldemar Fenn Deitrich.   1928, 383 pp., 70 photographs, 12 line cuts
including maps, cloth _____   1.75

Bulletin No. 100.   California Mineral Production for 1926, by Walter W.
Bradley, 1927, 174 pp., paper_____   .25

Bulletin No. 101.   California Mineral Production for 1927, by Henry H.
Symons.   1928, 311 pp., paper_____   .25

Bulletin No. 102.   California Mineral Production for 1928, by Henry H.
Symons.   1929, 210 pp., paper_____   .25

Bulletin No. 103.   California Mineral Production for 1929, by Henry H.
Symons, 1930.   231 pp., paper_____   .25

Bulletin No. 104.   Bibliography of the Geology and Mineral Resources of
California, to the end of 1930, by Solon Shedd_____   2.25

**Bulletin No. 105.   Mineral Production in California for 1930 and Direc-
tory of Producers _____   ____

Bulletin No. 106.   Manner of Locating and Holding Mineral Claims in
California (with forms)_____   .25

Bulletin No. 107.   Mineral Production in California for 1931 and Directory of Producers_____   .50

Bulletin No. 108.   Mother Lode Gold Belt of California, by Clarence A.
Logan, 1934, 240 pp., with geologic and claim maps, cloth_____   2.00

Bulletin No. 109.   California Mineral Production and Directory of Mineral Producers for 1932, by Henry H. Symons, 200 pp., paper_____   .25

Bulletin No. 110.   California Mineral Production and Directory of Mineral Producers for 1933, by Henry H. Symons, 214 pp., paper_____   .25

Bulletin No. 111.   California Mineral Production and Directory of Mineral Producers for 1934, by Henry H. Symons, 334 pp., paper_____   .75

Bulletin No. 112.   California Mineral Production and Directory of Mineral Producers for 1935, by Henry H. Symons, 205 pp., paper_____   .80

PRELIMINARY REPORTS

Asterisks (**) indicate the publication is out of print.

Price
Postpaid

**Preliminary Report No. 1. Notes on Damage by Water in California Oil
Fields, December, 1913. By R. P. McLaughlin, 4 pp_____ ____
**Preliminary Report No. 2. Notes on Damage by Water in California Oil
Fields, March, 1914. By R. P. McLaughlin, 4 pp_____ ____
Preliminary Report No. 3. Manganese and Chromium, 1917. By E. S.
Boalich. 32 pp._____ $0.05
**Preliminary Report No. 4. Tungsten, Molybdenum and Vanadium. By
E. S. Boalich and W. O. Castello, 1918. 34 pp. Paper_____ ____
**Preliminary Report No. 5. Antimony, Graphite, Nickel, Potash, Strontium
and Tin. By E. S. Boalich and W. O. Castello, 1918. 44 pp. Paper ____
Preliminary Report No. 6. A Review of Mining in California During
1919. By Fletcher Hamilton, 1920. 43 pp. Paper_____ .05
**Preliminary Report No. 7. The Clay Industry in California. By E. S.
Boalich, W. O. Castello, E. Huguenin, C. A. Logan, and W. B.
Tucker, 1920. 102 pp. 24 illustrations. Paper_____ ____
**Preliminary Report No. 8. A Review of Mining in California During
1921, with Notes on the Outlook for 1922. By Fletcher Hamilton,
1922. 68 pp. Paper_____ ____

MISCELLANEOUS PUBLICATIONS

**First Annual Catalogue of the State Museum of California, being the
collection made by the State Mining Bureau during the year ending
April 16, 1881. 350 pp._____ ____
**Catalogue of books, maps, lithographs, photographs, etc., in the library of
the State Mining Bureau at San Francisco, May 15, 1884. 19 pp.__ ____
**Catalogue of the State Museum of California, Volume II, being the col-
lection made by the State Mining Bureau from April 16, 1881, to
May 5, 1884. 220 pp._____ ____
**Catalogue of the State Museum of California, Volume III, being the col-
lection made by the State Mining Bureau from May 15, 1884, to
March 31, 1887. 195 pp._____ ____
**Catalogue of the State Museum of California, Volume IV, being the col-
lection made by the State Mining Bureau from March 30, 1887, to
August 20, 1890. 261 pp._____ ____
**Catalogue of the Library of the California State Mining Bureau, Sep-
tember 1, 1892. 149 pp._____ ____
**Catalogue of West North American and Many Foreign Shells with Their
Geographical Ranges, by J. G. Cooper. Printed for the State Mining
Bureau, April, 1894 _____ ____
**Report of the Board of Trustees for the four years ending September,
1900. 15 pp. Paper_____ ____
Bulletin. Reconnaisance of the Colorado Desert Mining District. By
Stephen Bowers, 1901. 19 pp. 2 illustrations. Paper_____ .10
Commercial Mineral Notes. A monthly mimeographed sheet, beginning
April, 1923 _____(15c annually) Free

MAPS

Register of Mines with Maps

**Register of Mines, with Map, Amador County _____ ____
Register of Mines, with Map, Butte County _____ .30
**Register of Mines, with Map, Calaveras County _____ ____
**Register of Mines, with Map, El Dorado County_____ ____
**Register of Mines, with Map, Inyo County _____ ____
**Register of Mines, with Map, Kern County _____ ____
**Register of Mines, with Map, Lake County _____ ____
**Register of Mines, with Map, Mariposa County _____ ____
Register of Mines, with Map, Nevada County _____ .30
**Register of Mines, with Map, Placer County _____ ____
**Register of Mines, with Map, Plumas County _____ ____
**Register of Mines, with Map, San Bernardino County_____ ____

MAPS—Continued

Register of Mines with Maps

Asterisks (**) indicate the publication is out of print.

	Price Postpaid
Register of Mines, with Map, San Diego County	$0.30
Register of Mines, with Map, Santa Barbara County (1906)	.30
**Register of Mines, with Map, Shasta County	____
**Register of Mines, with Map, Sierra County	____
**Register of Mines, with Map, Siskiyou County	____
**Register of Mines, with Map, Trinity County	____
**Register of Mines, with Map, Tuolumne County	____
Register of Mines, with Map, Yuba County (1905)	.30
Register of Oil Wells, with Map, Los Angeles City (1906)	.10

OTHER MAPS

**Map of California, Showing Mineral Deposits (50x60 in.)	____
**Map of Forest Reserves in California	____
**Mineral and Relief Map of California	____
**Map of El Dorado County, Showing Boundaries, National Forests	____
**Map of Madera County, Showing Boundaries, National Forests	____
**Map of Placer County, Showing Boundaries, National Fortsts	____
**Map of Shasta County, Showing Boundaries, National Forests	____
**Map of Sierra County, Showing Boundaries, National Forests	____
**Map of Siskiyou County, Showing Boundaries, National Forests	____
**Map of Tuolumne County, Showing Boundaries, National Forests	____
**Map of Mother Lode Region	____
**Map of Desert Region of Southern California	____
Map of Minaret District, Madera County	.25
Map of Copper Deposits in California	.05
**Map of Calaveras County	____
**Map of Plumas County	____
**Map of Trinity County	____
**Map of Tuolumne County	____
**Geographical Map of Inyo County. Scale 1 inch equals 4 miles	____
**Map of California accompanying Bulletin No. 89, showing generalized classification of land with regard to oil possibilities. Map only, without Bulletin	____
Geological Map of California, 1916. Scale 1 inch equals 12 miles. As accurate and up-to-date as available data will permit as regards topography and geography. Shows railroads, highways, post offices and other towns. First geological map that has been available since 1892, and shows geology of entire state as no other map does. Geological details lithographed in 23 colors. Mounted	2.75
**Topographic Map of Sierra Nevada Gold Belt, showing distribution of auriferous gravels, accompanying Bulletin No. 92. In 4 colors (also sold singly)	____
Geologic Map of Northern Sierra Nevada, showing Tertiary River Channels and Mother Lode Belt accompanying July-October Chapter of Report XXVIII of the State Mineralogist. (Sold singly)	.25
Map of Northern California, showing rivers and creeks which produced placer gold in 1932	.20
Mother Lode Geologic and claim maps in 5 county sections: El Dorado, Amador, Calaveras, Tuolumne and Mariposa. Single sections 10c. Set of 5	.50
Map of Mariposa County, showing principal gold mines	.10
Geologic Map of Elizabeth Lake Quadrangle, Los Angeles and Kern Counties (accompanying October Chapter of Report XXX), sold separately	.10
Map of Western Portion of Siskiyou County Showing Location of Principal Gold Mines (accompanying July Chapter of Report XXXI), sold separately	.10
Geologic Map of Redding and Weaverville Quadrangles Showing Location of Gold Mines	.25
Map of Ancient Channel System, Calaveras County	.10
Map of Ancient Channels Between San Andreas and Mokelumne Hill	.10

OIL FIELD MAPS

The maps are revised from time to time as development work advances and ownerships change.

Price
(including postage)

Map No. 1—Sargent, Santa Clara County	$0.50
Map No. 2—Santa Maria, including Cat Canyon and Los Alamos	1.00
Map No. 3—Santa Maria, including Casmalia and Lompoc	1.00
Map No. 4—Brea Olinda and (East Portion) Coyote Hills, Los Angeles and Orange Counties	1.00
Map No. 6—Salt Lake-Beverly Hills, Los Angeles County	1.00
Map No. 7—Sunset and San Emidio, Kern County	1.00
Map No. 8—South Midway and Buena Vista Hills, Kern County	1.00
Map No. 9—North Midway and McKittrick, Kern County	1.00
Map No. 10—Belridge and McKittrick Front, Kern County	1.00
Map No. 11—Lost Hills and North Belridge, Kern County	1.00
Map No. 12—Devils Den, Kern County	.75
Map No. 13—Kern River, Kern County	1.00
Map No. 14—Coalinga, Fresno County	1.25
Map No. 15—Elk Hills, Kern County	1.00
Map No. 16—Ventura-Ojai, Ventura County	1.00
Map No. 17—Santa Paula-Sespe, including Bardsdale, South Mountain and Camarillo, Ventura County	1.00
Map No. 18—Piru-Simi-Newhall, Ventura County	1.00
Map No. 19—Arroyo Grande, San Luis Obispo County	.75
Map No. 20—Long Beach, Los Angeles County	1.50
Map No. 21-B—Portion of District No. 5, showing boundaries of oil fields—Fresno, Kings and Kern Counties	.75
Map No. 21-C—Portion of District No. 4, showing boundaries of oil fields—Kern, Kings and Tulare Counties	.75
Map No. 22—Portion of District No. 3, showing boundaries of oil fields—Santa Barbara County	.50
Map No. 23—Portion of District No. 2, showing boundaries of oil fields—Ventura County	.75
Map No. 24—Portion of District No. 1, showing boundaries of oil fields—Los Angeles and Orange Counties	.75
Map No. 26—Huntington Beach, Orange County	1.25
Map No. 27—Santa Fe Springs, Los Angeles County	1.00
Map No. 28—Torrance, Los Angeles County	1.25
Map No. 29—Dominguez, Los Angeles County	.75
Map No. 30—Rosecrans, Los Angeles County	1.00
Map No. 31—Inglewood, Los Angeles County	1.00
Map No. 32—Seal Beach, Los Angeles and Orange Counties	1.00
Map No. 33—Rincon, Ventura County	1.25
Map No. 34—Mt. Poso, Kern County	.75
Map No. 35—Round Mountain, Kern County	.75
Map No. 36—Kettleman Hills, Fresno, Kings and Kern Counties	1.25
Map No. 37—Montebello, Los Angeles County	75
Map No. 38—Whittier, Los Angeles County	1.00
Map No. 39—West Coyote, Los Angeles and Orange Counties	1.00
Map No. 40—Elwood, Santa Barbara County	1.00
Map No. 41—Potrero, Los Angeles County	.75
Map No. 42—Playa del Rey, Los Angeles County	1.25
Map No. 43—Capitan, Santa Barbara County	.75
Map No. 44—Mesa, Santa Barbara County	1.25
Map No. 45—Buttonwillow gas, Kern County	.75
Map No. 46—Richfield, Orange County	1.25
Map No. 48—Mountain View and Edison, Kern County	1.00
Map No. 49—Fruitvale, Kern County	1.00
Map No. 50—Wilmington, Los Angeles County	1.00
Map No. 51—Santa Maria Valley, Santa Barbara County	1.00
Map No. 52—El Segundo and Lawndale, Los Angeles County	1.50
Map No. 53—Greeley and Ten Section, Kern County	.75

DETERMINATION OF MINERAL SAMPLES

Samples (limited to two at one time) of any mineral found in the State may be sent to the Division of Mines for identification, and the same will be classified free of charge. No samples will be determined if received from points outside the State. It must be understood that no assays, or quantitative determinations will be made. Samples should be in lump form if possible, and marked plainly with name of sender on outside of package, etc. No samples will be received unless delivery charges are prepaid. A letter should accompany sample, giving locality where mineral was found and the nature of the information desired.